Memoirs of the American Mathematical Society

Number 431

Thomas G. Goodwillie

A multiple disjunction lemma for smooth concordance embeddings

Published by the

AMERICAN MATHEMATICAL SOCIETY

Providence, Rhode Island, USA

July 1990 · Volume 86 · Number 431 (first of 2 numbers)

1980 *Mathematics Subject Classification* (1985 *Revision*).
Primary 57.

Library of Congress Cataloging-in-Publication Data

Goodwillie, Thomas G. (Thomas Gehret), 1954–
A multiple disjunction lemma for smooth concordance embeddings/Thomas G. Goodwillie.
p. cm. – (Memoirs of the American Mathematical Society, ISSN 0065-9266; no. 431)
"July 1990."
Includes bibliographical references.
"Volume 86 number 431."
ISBN 0-8218-2493-7
1. Concordances (Topology) 2. Embeddings (Mathematics) 3. Piecewise linear topology.
I. Title. II. Series.
QA3.A57 no. 431
[QA613.4]
510 s–dc20 90-31826
[514] CIP

Subscriptions and orders for publications of the American Mathematical Society should be addressed to American Mathematical Society, Box 1571, Annex Station, Providence, RI 02901-1571. *All orders must be accompanied by payment.* Other correspondence should be addressed to Box 6248, Providence, RI 02940-6248.

SUBSCRIPTION INFORMATION. The 1990 subscription begins with Number 419 and consists of six mailings, each containing one or more numbers. Subscription prices for 1990 are $252 list, $202 institutional member. A late charge of 10% of the subscription price will be imposed on orders received from nonmembers after January 1 of the subscription year. Subscribers outside the United States and India must pay a postage surcharge of $25; subscribers in India must pay a postage surcharge of $43. Each number may be ordered separately; *please specify number* when ordering an individual number. For prices and titles of recently released numbers, see the New Publications sections of the NOTICES of the American Mathematical Society.

BACK NUMBER INFORMATION. For back issues see the AMS Catalogue of Publications.

MEMOIRS of the American Mathematical Society (ISSN 0065-9266) is published bimonthly (each volume consisting usually of more than one number) by the American Mathematical Society at 201 Charles Street, Providence, Rhode Island 02904-2213. Second Class postage paid at Providence, Rhode Island 02940-6248. Postmaster: Send address changes to Memoirs of the American Mathematical Society, American Mathematical Society, Box 6248, Providence, RI 02940-6248.

Printed in the United States of America.
The paper used in this book is acid-free and falls within the guidelines established to ensure permanence and durability. ∞

10 9 8 7 6 5 4 3 2 1 95 94 93 92 91 90

TABLE OF CONTENTS

Abstract. We prove the following theorem in concordance theory, generalizing a result of Morlet. Denote by $C(P,N)$ the space of smooth concordance embeddings of P in N, if N is a smooth manifold and $P \subset N$ is a smooth proper compact submanifold.

THEOREM. Let N^n be a smooth manifold, and let P^p, $Q_1^{q_1}, \cdots Q_a^{q_a}$ $(a \geq 1)$ be disjoint smooth proper compact submanifolds of N such that the codimensions $n-p$ and $n-q_j$ are all at least three. Then the homotopy groups of the $(a+1)$-ad

$$(C(P,N);\ C(P,N-Q_1), \cdots C(P,N-Q_a))$$

vanish in dimensions less than or equal to $n-p-3 + \sum_{j=1}^{a} (n-q_j-1)$.

The method is a differentiable analogue of the piecewise linear technique called "sunny collapsing". Given any smooth parametrized family of concordance embeddings (representing a homotopy class), we associate to it a hierarchy of singularity sets. This hierarchy is such that if we can get all singularities "out of the way" then we can prove that the homotopy class is trivial. On the other hand for some singularity types an inductive construction is possible; if all singularities of "worse than" the given type are already out of the way, then those of that type can be dealt with. The construction fails for some kinds of singularities, and this determines the range of dimensions occurring in the statement

Key Words: pseudoisotopy, concordance, embedding, disjunction, excision

Received by Editor September 5, 1986.

To my parents

Introduction

This is a slightly revised version of the author's Ph.D. thesis (1982, Princeton). The main result is a generalization of Morlet's Disjunction Lemma. This introduction is organized as follows:

§A states the basic definitions and problems of concordance theory.

§B is a summary of previously known results on concordance embeddings, including Morlet's lemma.

§C states our main result and gives some indication of how it can be used.

§D is a long, sketchy discussion of the method used in proving Theorem D. This discussion may make the actual proof, which occupies the remainder of the thesis, a little easier to read.

I would like to thank my thesis adviser Wu-chung Hsiang for all his help and encouragement. I would also like to thank John Mather for an extremely helpful conversation.

§Intro. A. Spaces of Concordances.

A great deal of recent work in the topology of manifolds has been aimed at trying to understand the homotopy type of the group of automorphisms of a manifold. For good reasons most of this work has dealt with spaces of concordances. We quickly recall how these are defined.

Let N be a compact manifold*. A concordance of N is a diffeomorphism

$$I \times N \xrightarrow{F} I \times N ,$$

where $I = [0,1]$, such that

(1) $$F\Big|_{0\times N \cup I\times\partial N} = \text{inclusion} .$$

The space of all such F (with the C^∞ topology) is denoted $C(N)$. The space $C(N)$ has been studied by a number of methods (see [Hat] for a survey of results up to 1976). The general aim of all of these is to determine to what extent the homotopy type of $C(N)$ is a function of the

---

*All manifolds and maps are smooth $(= C^\infty)$ unless we say otherwise. Manifolds may have boundaries and corners.

homotopy type of N and to understand this function as well as possible.

One useful tool in studying concordances is the notion of concordance embedding. If $P \subset N$ is a compact submanifold then a concordance embedding of P in N is an embedding

$$I \times P \xrightarrow{F} I \times N$$

such that

$$(2) \qquad \begin{cases} F\big|_{0 \times P \,\cup\, I \times (P \cap \partial N)} = \text{inclusion} \\ F(1 \times P) \subset 1 \times N \\ F \text{ is transverse to boundary and corner sets} \end{cases}$$

The space of all such F is denoted $C(P,N)$. Information about $C(P,N)$ is relative information about $C(N)$, because up to homotopy there is a fibration

$$C(N') \longrightarrow C(N) \longrightarrow C(P,N) ,$$

where $N' \subset N$ is the closed complement of a regular neighborhood of P in N.

The objects of study in this thesis are spaces $C(P^p,N^n)$ where the codimension $n-p$ is at least three. Thus, roughly speaking, we are investigating that "part" of $C(N)$ which does not "come from" the fundamental group of N. The statement that we obtain (Theorem D below) is quite technical. We will use it in a future paper to determine the "stable range" for $C(P,N)$ if $n-p\geq 3$ (i.e., the range of dimensions in which its homotopy type only depends on the homotopy type of the pair $(N,N-P)$), and to obtain, in this range, a description of $C(P,N)$ in terms of classical homotopy theory. (See §C for further details.) This contrasts sharply with results concerning the effect of the fundamental group of N on $C(N)$ ([HW], [I1], [I2]), where the computations always involve algebraic K-theory.

§Intro. B. Known Results .

From now on "concordance" will mean "concordance embedding", and embeddings will have codimension ≥ 3. The first result in the subject is Hudson's "Concordance-Implies-Isotopy Theorem", which can be stated as follows:

THEOREM A. (See, e.g., [Hu], Theorem 2, p.12). Let N^n be a smooth manifold, $P^p \subset N^n$ a compact proper submanifold with $n-p \geq 3$. Then $C(P,N)$ is connected.

Morlet improved considerably on Theorem A. His main technical result in the subject is the following "Disjunction Lemma" ([BLR], p.1).

THEOREM B. (Morlet) Let N^n be a smooth manifold, P^p and Q^q compact proper submanifolds of N, $P \cap Q = \phi$, $n-p \geq 3$, $n-q \geq 3$. Then the pair $(C(P,N), C(P,N-Q))$ is $(2n-p-q-4)$ - connected.

REMARK 1. This is not exactly the result of Morlet as given in [BLR]. For one thing, Morlet only deals with the case in which P and Q are disks (which is in fact the most important case). For another, he only asserts the pair to be $(2n-p-q-5)$ - connected, except in the special case in which N is simply connected. Finally, he works simultaneously in the smooth and PL categories. The slightly improved version given above (in the smooth case) is proved in this thesis; it is a special case of Theorem D below.

REMARK 2. In the PL case Millett ([Mi1]; [Mi2], p. 366) has given an alternative proof of Morlet's Lemma using "fibered sunny collapsing." He obtains the same improvement by one dimension which we obtain in the smooth case. This is no coincidence; the technique which we use can be viewed as a smooth version of sunny collapsing.

The usefulness of Theorem B comes largely from the following "delooping trick". (See [BLR] pp. 23-25). Let $D^p \subset N^n$ be a properly embedded disk, with $n-p\geq 3$. Consider the smaller disk $D^{p-1} \subset D^p$ (embedded in the usual way). D^{p-1} cuts D^p into two closed half-disks D^p_+ and D^p_-. An elementary argument shows that $C(D^p_+,N)$ is contractible. Also $C(D^p_+,N)$ fibers over $C(D^{p-1},N)$ by $F \longrightarrow F\big|_{I\times D^{p-1}}$.

(It fibers over a union of path-components by the isotopy extension theorem, and $C(D^{p-1},N)$ is connected by Theorem A.) The fiber of this map can be identified (up to homotopy type) with $C(D^p, N \cup h)$, where h is a handle of index $n-p$ attached to $N-D^p$ according to the following picture:

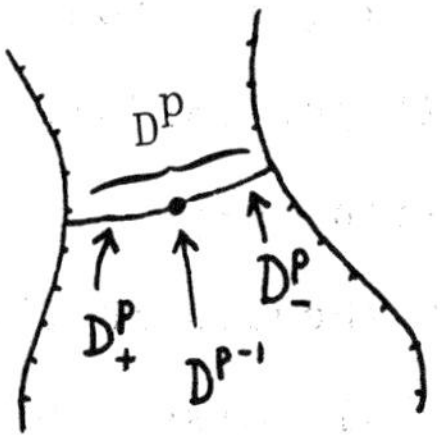

N

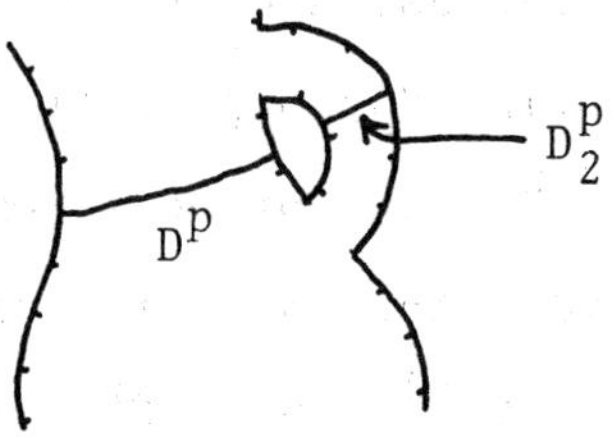

N ∪ h

The reason is roughly that the second picture above is isomorphic to

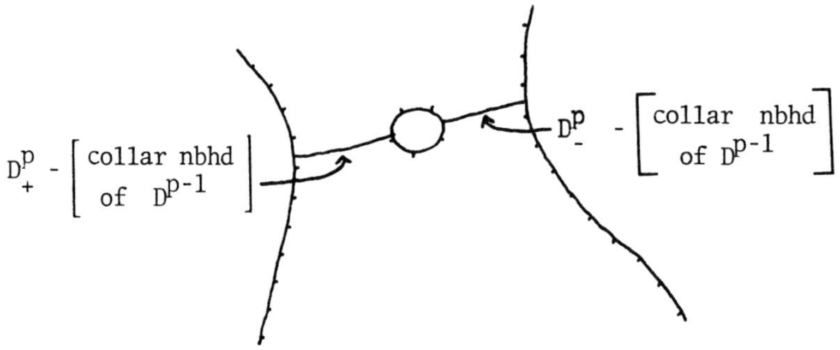

N-[Tubular neighborhood of D^{p-1}]

Therefore $C(D^p, N \cup h) \simeq \Omega\, C(D^{p-1},N)$.

Let D^p_2 be the cocore of h. Thus $C(D^p, N \cup h - D^p_2) \simeq C(D^p, N)$. Theorem B says that the pair

$$(C(D^p, N \cup h), C(D^p, N \cup h - D^p_2))$$

is $(2n-2p-4)$-connected. Therefore we have a $(2n-2p-4)$-connected map (well-defined up to homotopy)

$$C(D^p, N) \longrightarrow \Omega C(D^{p-1}), N) .$$

Iterating this construction we get a $(2n-2p-4)$-connected map

$$C(D^p, N) \longrightarrow \Omega^p C(D^o, N^n) .$$

Now, the space $C(D^o, N^n)$ is easier to understand than $C(D^p, N)$. In particular, an easy general position argument shows that it is $(n-3)$-connected, so that we have an improved form of Theorem A:

THEOREM C. $C(D^p, N^n)$ is $(n-p-3)$-connected.

In fact, it is not hard to do better than this; there is a map

$$C(D^o, N^n) \longrightarrow \Omega^\infty S^\infty S^{n-2} \Omega N$$

which can be shown to be $(2n-5)$-connected ([Go1],Lemma 7.12). Thus in the range $0 \le i \le 2n-2p-5$ we have

$$\pi_i C(D^p, N^n) \cong \pi^S_{i+2-n}(\Omega N) .$$

To go beyond this range we need "multiple disjunction".

§Intro. C. The Multiple Disjunction Lemma.

The computation of $\pi_i C(D^p, N^n)$ for $i < 2n-2p-4$ relies, as we have just seen, on a vanishing theorem (Theorem B) for the first few relative homotopy groups of a "disjunction pair" of concordance spaces. To get information about higher values of i it seems worthwhile to try to compute the first few homotopy groups of a "disjunction pair" beyond the range of applicability of Theorem B. Let us try to do this by applying to a pair $(C(D^p, N), C(D^p, N-Q))$ the same "delooping trick" which we have already applied to a single space $C(D^p, N)$.

We want to show that the map of pairs

$$(C(D^p, N), C(D^p, N-Q)) \longrightarrow$$
$$(\Omega C(D^{p-1}, N), \Omega C(D^{p-1}, N-Q))$$

induces isomorphisms of relative homotopy groups in some range of low dimensions. By the delooping trick it is enough to show that the same is true for the inclusion of pairs

$$(C(D^p,N),\ C(D^p,N-Q)) \longrightarrow$$

$$(C(D^p,N \cup h),\ C(D^p,N \cup h-Q)).$$

where h is an $(n-p)$-handle attached to $N-D^p-Q$ as before. In other words, we want a vanishing theorem for low-dimensional homotopy groups of the triad

$$(C(D^p,N \cup h)\ ;\ C(D^p,N \cup h-D_2^p),C(D^p,N \cup h-Q)).$$

It will turn out that any triad of the form

$$(C(P^p,N^n)\ ;\ C(P^p,N^n-Q_1^{q_1}),\ C(P^p,N^n-Q_2^{q_2}))$$

where P, Q_1, and Q_2 are disjoint compact proper submanifolds of N with codimensions $n-p$, $n-q_1$, and $n-q_2$ all at least three, has trivial triad homotopy groups in dimensions $\leq 3n-p-q_1-q_2-5$. More generally, consider "disjunction $(a+1)$-ads." These are $(a+1)$-ads*

$$C = (C(P^p,N^n)\ ;\ C(P^p,N-Q_1^{q_1}),\cdots,C(P^p,N-Q_a^{q_a}))$$

where P, $Q_1,\cdots Q_a$ are disjoint compact proper submanifolds of N. Our main result is:

---

*See §I.B. for definitions.

THEOREM D. (Multiple Disjunction Lemma). Let C be a disjunction $(a+1)$-ad as above, and assume $a \geq 1$, $n-p \geq 3$, $n-q_j \geq 3$ $\forall_j$. Then the homotopy groups $\pi_i(C)$ vanish for

$$i \leq n-p-3+\sum_{j=1}^{a}(n-q_j-1) .$$

Theorem D combined with the delooping trick shows that the natural map of a-ads

$$\left(C(D^p,N^n)\ ;\ \left\{C(D^p,N^n-Q_j^{q_j})\right\}_{j=1}^{a-1}\right) \longrightarrow$$

$$\left(\Omega C(D^{p-1},N^n)\ ;\ \left\{\Omega\ C(D^{p-1},N^n-Q_j^{q_j})\right\}_{j=1}^{a-1}\right)$$

will, under the usual assumptions, induce isomorphisms of homotopy groups in dimensions less than

$$2n-2p-4+\sum_{j=1}^{a-1}(n-q_j-1)$$

and an epimorphism in this dimension. This provides a useful device for making inductive arguments with respect to p.

We hope to show in a future paper that for $n-p\geq 3$ the homotopy type of $C(P,N)$ can be completely described in terms of the spaces $C(F,N)$ where F ranges over the finite subsets of P. From this description will follow:

(3) A stability theorem: The suspension map (see [BL] p.451) from $C(P,N)$ to $C(P\times I, N\times I)$ is $(2n-p-4)$-connected, if $n-p \geq 3$.

(4) A spectral sequence $E^r_{i,j} \Rightarrow \pi_{i+j}C(P,N)$ in which:

$$E^2_{i,j} \cong \pi_{i+j}(\text{the space of } S_j\text{-equivariant compactly supported sections of a certain bundle over } \underbrace{P\times P\times\cdots\times P}_{j} - \text{diagonal})$$

$$E^2_{i,j} = 0 \quad \text{if} \quad j<1 \quad \text{or} \quad i<(n-p-3)j$$

In the stable range $(i+j < 2n-p-4)$ $E^2_{i,j}$ is computable in terms of homotopy theory.

Set $C(N) = \varinjlim_k C(N\times I^k)$, a direct limit under the suspension map, and make C a homotopy functor of finite CW complexes (like the functor C_{Diff} of [Hat]) or better yet a homotopy functor of pairs of countable CW complexes. Then (4) becomes a "computation" of $\pi_* C(X,Y)$ for any 2-connected pair (X,Y).

In particular take X to be contractible and Y to be a 1-connected pointed space, and tensor with the rational numbers $\mathbb{Q}$. Our E^2-term becomes

$$E^2_{i,j} \otimes \mathbb{Q} \cong \tilde{H}_{i+3j}(\underbrace{SY \wedge \cdots SY}_{j} ; \mathbb{Q})^{\mathbb{Z}_j} , \quad j > 1$$

(the $\mathbb{Z}_j$-invariant subgroup of the reduced rational homology of a j-fold smash product of copies of the suspension of Y ; $\mathbb{Z}_j$ acts by permuting the factors in the smash product). Quite similar results have been obtained by Dwyer-Hsiang-Staffeldt ([DHS1], [DHS2], [HS1], [HS2]) using very different methods; they compute the reduced rational A-theory of a simply-connected space, where A is Waldhausen's "algebraic K-theory of topological spaces." ([W1], [W2]). As the simplest example we confirm their result ([HS1]) that for $n \geq 2$

(5) $$\pi_\ell C(S^n) \otimes \mathbb{Q} \cong$$

$$(\pi_\ell C(\text{point}) \otimes \mathbb{Q}) \oplus \begin{cases} \mathbb{Q} & \text{if } \ell+1 = j(n-1),\ j \geq 2, \\ & \text{and either } j \text{ or } n \text{ odd} \\ 0 & \text{otherwise.} \end{cases}$$

Our method also applies in the non-simply connected case, and there even the rational results are new. For example, for any group π and $n\geq 2$ we have:

$$(6) \qquad \pi_\ell C(S^n \vee K(\pi,1)) \otimes \mathbb{Q} \cong$$

$$(\pi_\ell C(K(\pi,1)) \otimes \mathbb{Q}) \oplus \begin{cases} H^0(\mathbb{Z}_j ; \mathbb{Z}^\varepsilon \otimes \mathbb{Q}[\overbrace{\pi\times\cdots\times\pi}^{j \text{ times}}]) \\ \quad \text{if } \ell+1 = j(n-1),\ j\geq 2 \\ \mathbb{Q}[\pi]/\mathbb{Q} \quad \text{if } \ell+1 = n-1 \\ 0 \quad \text{otherwise} \end{cases}$$

where $\mathbb{Z}^\varepsilon$ is $\mathbb{Z}$ with the nontrivial $\mathbb{Z}_j$-action if $j \equiv n \equiv 0 \pmod 2$ and the trivial $\mathbb{Z}_j$-action otherwise, and $\mathbb{Q}[\pi\times\cdots\times\pi]$ is the $\mathbb{Z}_j$-module given by the permutation action of $\mathbb{Z}_j$ on the set $\pi\times\cdots\times\pi$.

(Added in revision: Unfortunately the proof of (3) has not yet been written down, and that of (4) might never be. On the other hand the idea behind (4) has borne other fruit: There is now an algebraic proof ([Go3]) of relative rational statements like (6)--or rather, of corresponding statements about K-theory which by Waldhausen's work imply statements about concordance theory. There has also been quite a lot of work ([CCGH],[O],[Go1],[Go2],[BCCGHM]) on relative Waldhausen K-theory (not just its rationalization).)

§Intro. D. Sketch of the Proof.

In order to make the proof of Theorem D more readable we will devote the rest of the introduction to a leisurely discussion of how one might prove Theorem B (the special case of D which is essentially Morlet's Lemma). The proof sketched here is not quite the same as the one actually given in Chapters I-III. The sketched proof uses certain sophisticated notions which we have chosen to avoid in the actual proof, namely semi-algebraic sets, stratifications of infinite-dimensional manifolds, and the Hilbert scheme (of zero-dimensional varieties in an algebraic variety). By avoiding these we have made the proof more elementary, but also perhaps conceptually less clear — for example, we avoid semi-algebraic sets by doing our algebraic geometry over the complex numbers. The following discussion is intended to compensate partially for that loss of clarity. During this discussion we will of course shamelessly ignore many technical details.

§Intro. D.1. The overall plan.

Let N^n be a (smooth) manifold with disjoint proper compact submanifolds P^p and Q^q, and assume $n-p\geq 3$, $n-q\geq 3$. We have to show that the pair

$$(C,C') \underset{\text{def}}{=} (C(P,N),\ C(P,N-Q))$$

is $(2n-p-q-4)$-connected. That is, if $0\leq s\leq 2n-p-q-4$ we must show that every continuous map of pairs

$$(D^s,\partial D^s) \longrightarrow (C,C')$$

is homotopic as a map of pairs to a map

$$(D^s,\partial D^s) \longrightarrow (C',C').$$

We may restrict our attention to "smooth" maps, that is, maps

$$\begin{aligned} D^s &\longrightarrow C \\ y &\longmapsto F_y \end{aligned}$$

such that the associated map

$$I \times P \times D^s \xrightarrow{F} I \times N \times D^s$$

$$(t,x,y) \longrightarrow (F_y(t,x),y)$$

is a smooth embedding (and ∂D^s is mapped into C').

The plan is to make a stratification* $\mathcal{S} = \{\Sigma\}$ of $P \times C$, thinking of C as an infinite-dimensional manifold, such that:

(7) the strata are of two kinds, called "good" and "bad";

(8) each bad stratum Σ has codimension $\geq n-2$, so that if F is in general position in a suitable sense with respect to Σ and Q then we have

$(9)_\Sigma$ $$(x,F_y) \in \Sigma \Rightarrow F_y(I \times x) \cap Q = \phi ;$$

and

---

*That is, a collection of finite-codimensional submanifolds Σ (the strata) such that the complement of the union of the strata has infinite codimension, there are only finitely many strata of any codimension, and the union of the strata of codimension less than K is open for all K.

(10) for each good stratum Σ, an inductive construction is possible: If F satisfies $(9)_{\Sigma'}$ for all strata Σ' of greater codimension than Σ, then we can alter the <u>fibered concordance</u> F by a <u>fibered isotopy of concordances</u> (see §I.A.1. for definitions) to make F satisfy $(9)_\Sigma$ as well.

At the end of the induction we will have $(9)_\Sigma$ for all Σ, and thus $F_y \in C'$ for all $y \in D^s$, as required.

We will describe the stratification in §D.2, §D.3, and §D.4. In §D.5 we will try to explain (7), (8), and (10).

§Intro. D.2. A stratification.

We begin constructing S. Consider the set

$$S_o \overset{=}{\text{def}} \left\{ (x_1,x_2,F) \in (I\times P)^{(2)}\times C \mid Fx_1 \text{ is below } Fx_2 \right\}$$

(Here $(I\times P)^{(r)}$ is the space of ordered r-tuples of distinct points in $I\times P$, and we say that (t_1,x_1) is below (t_2,x_2) in $I\times N$ if $t_1 < t_2$ and $x_1 = x_2$.)

The set S_o is an obvious one to consider, for the following reason: Suppose one wants to move a concordance

$$I\times P \xrightarrow{F=(h,f)} I\times N$$

through an isotopy F^u of concordances so as to get a new concordance F^1 which is "level-preserving", i.e., $F^1(t,x) = (t,f^1(t,x))$. The crudest attempt at this is the homotopy

$$I\times P \xrightarrow{F^u} I\times N, \quad 0\le u\le 1$$

$$(t,x) \longmapsto ((1-u)h(t,x) + ut, f(t,x)).$$

The attempt fails if some F^u is not injective; and if this happens it is because S_o intersects $(I\times P)^{(2)} \times \{F\}$. (Making concordances level-preserving is as good as making them trivial, since the space of level-preserving concordances is contractible.)

S_o is a submanifold of codimension n in $(I\times P)^{(2)}\times C$. The first candidates for strata of S are the images $p\pi_1 S_o$ and $p\pi_2 S_o$, where π_i and p are the projections:

$$(I\times P)^{(r)}\times C \xrightarrow{\pi_i} I\times P\times C$$

$$(x_1,\cdots x_r,F)\longrightarrow (x_i,F)$$

$$I\times P\times C \xrightarrow{p} P\times C$$

$$(t,x,F)\longrightarrow (x,F)$$

§Intro. D.2.a. Operation B.

Unfortunately $p\pi_1|_{S_o}$ and $p\pi_2|_{S_o}$ are not embeddings; nor are their images disjoint. Wherever $p\pi_i|_{S_o}$ fails to embed or $p\pi_1|_{S_o}$ and $p\pi_2|_{S_o}$ have a common image point, we intend to relegate the image point to a stratum of higher codimension. What remains of $p\pi_1 S_o$ and $p\pi_2 S_o$ should then be two disjoint submanifolds of codimension $n-p-2$ in $P\times C$.

In order to detect these kinds of unpleasant behavior of the maps $p\pi_i|_{S_o}$ we introduce more sets like S_o. For example, let

$$S' = \left\{ (x_1,x_2,x_3,x_4,F) \in (I\times P)^{(4)} \times C \;\middle|\; (x_1,x_2,F) \in S_o,\ (x_3,x_4,F) \in S_o, p(x_2,F) = p(x_3,F) \right\}$$

$$S'' = \left\{ (x_1,x_2,x_3,F) \in (I\times P)^{(3)} \times C \;\middle|\; (x_1,x_2,F) \in S_o,\ (x_2,x_3,F) \in S_o \right\}$$

The set S' (resp. S'') is a submanifold of codimension $2n+p$ (resp. $2n$), so its images $p\pi_i\, S'$, $1\le i\le 4$ (resp. $p\pi_i\, S''$, $1\le i\le 3$) "ought" to be manifolds in $P\times C$ of codimension $(2n+p)-3(p+1)-1 = 2n-2p-4$ (resp. $2n-2(p+1)-1 = 2n-2p-3$). Of course, the images need not be manifolds at all, or be disjoint from each other; and so we are led to make still more sets to detect the intersections and self-intersections of the maps $p\pi_i\big|_{S'}$, $p\pi_i\big|_{S''}$.

To make this procedure systematic, we introduce an operation: Let S_α and $S_{\alpha'}$ be subsets of $(I\times P)^{(r_\alpha)} \times C$ and $(I\times P)^{(r_{\alpha'})} \times C$ respectively. Let

$$\{1,\cdots r_\alpha\} \xrightarrow{\phi} \{1,\cdots r\} \quad \text{and}$$

$$\{1,\cdots r_{\alpha'}\} \xrightarrow{\phi'} \{1,\cdots r\}$$

be injective maps (for some $r>0$) such that

$$\phi(\{1,\cdots r_\alpha\}) \cup \phi'(\{1,\cdots r_{\alpha'}\}) = \{1,\cdots r\} .$$

For each $i \in \{1,\cdots r_\alpha\}$ and $i' \in \{1,\cdots r_{\alpha'}\}$, set

$$B_{\phi,\phi',i,i'}(S_\alpha, S_{\alpha'}) = \left\{ x \in (I\times P)^{(r)} \times C \,\middle|\, \phi^*x \in S_\alpha ,\ \phi'^*x \in S_{\alpha'} ,\ p\pi_i\phi^*x = p\pi_{i'}\phi'^*x \right\} ,$$

where $\phi^*(x_1,\cdots x_{r_\alpha},F) = (x_{\phi(1)},\cdots x_{\phi(r_\alpha)},F)$.

In particular $S' = B_{\phi,\phi',2,1}(S_o, S_o)$, where

$$\{1,2\} \xrightarrow{\ \phi\ } \{1,2,3,4\}$$
$$1 \longmapsto 1$$
$$2 \longmapsto 2$$
$$\{1,2\} \xrightarrow{\ \phi'\ } \{1,2,3,4\}$$
$$1 \longmapsto 3$$
$$2 \longmapsto 4 ,$$

and $S'' = B_{\phi'',\phi''', 2, 1}(S_o, S_o)$, where

$$\{1,2\} \xrightarrow{\phi''} \{1,2,3\}$$

$$1 \longmapsto 1$$

$$2 \longmapsto 2$$

$$\{1,2\} \xrightarrow{\phi'''} \{1,2,3\}$$

$$1 \longmapsto 2$$

$$2 \longmapsto 3 \quad .$$

When Operation B is applied iteratively to S_0 in all possible ways it yields an infinite list of manifolds S_α lying in various sets $(I\times P)^{(r_\alpha)} \times C$. Inside each S_α, tentatively define W_α by:

$$W_\alpha = \left\{ x\in S_\alpha \mid \forall_i (1\le i\le r_\alpha) \;\; p\pi_i x \text{ does not lie in any set } p\pi_j S_\beta \text{ strictly smaller than } p\pi_i S_\alpha \right\}.$$

Each image $p\pi_i W_\alpha$ is now a candidate to be a stratum. Note that sometimes $p\pi_i S_\alpha = p\pi_j S_\beta$ for $(i,\alpha) \neq (j,\beta)$ (this redundancy will not bother us), but that otherwise the "strata" are disjoint and are injective images of the W_α.

§Intro. D.2.b. Operation C°.

However, we do not yet have a stratification. For one thing, we haven't made sure that the sets $p\pi_i W_\alpha$ are manifolds; $p\pi_i|_{W_\alpha}$ is an injection but not necessarily an immersion. Just as Operation B creates manifolds in which to put the self-intersection points of $p\pi_i|_{S_\alpha}$, we need another operation to handle any possible non-immersion points. So for each S_α and each i $(1 \le i \le r_\alpha)$ define

$$C_i^\circ(S_\alpha) = \left\{ x \in S_\alpha \,\middle|\, p\pi_i|_{S_\alpha} \text{ at } x \text{ is not an immersion} \right\} .$$

A new difficulty now arises: some of these new sets — $C_1^\circ(S_\alpha)$, for example — are not manifolds. In particular it is not clear that an expression such as $C_1^\circ C_1^\circ S_\alpha$ makes sense. We will shortly clear up this and other questions. First, however, we define three more operations A, D, and C.

§Intro. D.2.c. Operation A.

In order to relegate non-manifold points to lower strata we define

$$A(S_\alpha) = \left\{ x \in S_\alpha \,\middle|\, \begin{array}{l} \text{no neighborhood of } x \\ \text{in } S_\alpha \text{ is a manifold with} \\ \text{codimension } = \operatorname{codim}(S_\alpha, (I\times P)^{(r_\alpha)} \times C) \\ \text{in } (I\times P)^{(r_\alpha)} \times C \end{array} \right\} .$$

§Intro. D.2.d. Operation D.

An important matter which we have ignored up to now has to do with limit points of strata. If a limit point of $p\pi_i W_\alpha$ is not in $p\pi_i W_\alpha$ but is in $p\pi_i S_\alpha$ then by definition it must lie in some smaller $p\pi_j S_\beta$. If it is not in $p\pi_j W_\beta$ then it is in some still smaller $p\pi_k S_\gamma$, and so on. Eventually it is in some $p\pi_{i'} W_{\alpha'}$ of greater codimension than $p\pi_i S_\alpha$, or in a set of infinite codimension. However, $p\pi_i W_\alpha$ may have limit points outside of $p\pi_i S_\alpha$; $p\pi_i S_\alpha$ is not in general closed in $P\times C$. The map

$$(I\times P)^{r_\alpha} \times C \xrightarrow{p\pi_i} P\times C$$

is proper, but S_α is not in general closed in $(I\times P)^{r_\alpha} \times C$; it is only closed in $(I\times P)^{(r_\alpha)} \times C$.

We need another operation to gain control of any diagonal limit points which S_α may have. For any $S_\alpha \subset (I\times P)^{(r_\alpha)} \times C$ and any surjective but not bijective map

$$\{1,\cdots r_\alpha\} \xrightarrow{\ \phi\ } \{1,\cdots r\} ,$$

define

$$D_\phi(S_\alpha) = \left\{ x \in (I\times P)^{(r)} \times C \;\middle|\; \phi^* x \in \overline{S}_\alpha \right\}.$$

("Bar" denotes closure in $(I\times P)^{r_\alpha} \times C$.) For example let $\phi : \{1,2\} \to \{1\}$. Then

$$S_1 \underset{\text{def}}{=} D_\phi(S_o) = \left\{ (x,F) \in (I\times P) \times C \;\middle|\; \ker D(pF)_x \neq \{0\} \right\},$$

where p is the projection $I\times N \longrightarrow N$.

§Intro. D.2.e. Operation C.

Before going further we now define the last operation (Operation C). Notice that for any S_α and $i \in \{1,\cdots r_\alpha\}$ we have in our collection of sets

$$B_{id,\phi,i,1}(S_\alpha,S_1) = \left\{ (x_1,\cdots x_{r_\alpha},F) \in S_\alpha \;\middle|\; \ker D(pF)_{x_i} \neq \{0\} \right\},$$

where

$$\phi : \{1\} \longrightarrow \{1, \cdots r_\alpha\}$$
$$1 \longmapsto i \qquad .$$

This set $B_{\phi, id, i, 1}(S_\alpha, S_1)$ may be (Case 1) strictly smaller Than S_α or (Case 2) equal to it. In Case 1 for any $(x_1, \cdots x_{r_\alpha}, F) \in W_\alpha$ we have $\ker D(pF)_{x_i} = \{0\}$, by definition of W_α. In case 2 we have $\dim \ker D(pF)_{x_i} = 1$ for all $(x_1, \cdots x_{r_\alpha}, F) \in S_\alpha$, in particular in $W_\alpha^\cdot$. Thus in Case 2 there is a unique vector field $\xi_{\alpha,i}$ along $\pi_i|_{W_\alpha}$ such that

$$(DF) \cdot \xi_{\alpha,i} = \frac{\partial}{\partial t} \text{ identically.}$$

(Here $\frac{\partial}{\partial t}$ is the "unit upward vertical vector field" in $I \times N$.) Set $\eta_{\alpha,i} = (Dp) \cdot \xi_{\alpha,i}$, a vector field along $p\pi_i|_{W_\alpha}$.

Now, for reasons which it is too early to explain (we will come back to this subject at the very end of the Introduction) we will want (in Case 2, i.e. when $\eta_{\alpha,i}$ is defined) that

(11) $\eta_{\alpha,i}(x)$ is never a nonzero tangent to $p\pi_i W_\alpha$ at $p\pi_i(x)$, for $x \in W_\alpha$.

Therefore let $C_i(S_\alpha)$ be the closure in S_α of the set of all $(x,F) \in S_\alpha$ at which for some non-zero tangent vector ζ and some $c \in \mathbb{R}$ we have

$$D(pF)(D\pi_i\zeta - c\frac{\partial}{\partial t}) = 0$$

Thus $C_i(S_\alpha)$ contains $C_i^0(S_\alpha)$ and in Case 2 it also contains the points in W_α where (11) fails.

§Intro. D.2.f. The Stratification.

This, then, is the full collection of operations: A,B,C, and D. (C renders C° obsolete.) Starting with the set S_0 and repeatedly applying these in all possible ways, we get a countable collection of sets $S_\alpha \subset (I\times P)^{(r_\alpha)} \times C$. Order them according to

$$\text{(12)} \qquad c(S_\alpha) \underset{\text{def}}{=} \operatorname{codim}\,(S_\alpha,(I\times P)^{(r_\alpha)} \times C) - r_\alpha(p+1) \ .$$

Define

(13) $$W_\alpha = \left\{ x \in S_\alpha \,\middle|\, \forall i \in \{1,\dots r_\alpha\}\ p\pi_i x \text{ is not in any set } p\pi_j S_\beta \underset{\neq}{\subset} p\pi_i S_\alpha \right\}.$$

Then the sets $p\pi_i W_\alpha$ should be manifolds of codimension $c(S_\alpha) + p$ in $P\times C$, and should be the strata of a stratification of $P\times C$ (except for the "top stratum" of codimension zero, which we take to be $P\times C - \overline{\bigcup_{\alpha,i} p\pi_i S_\alpha}$).

§Intro. D.3. Why the stratification is a stratification.

In order to prove that this is a stratification (and even that it makes sense to apply the operations as indicated) we would have to do a certain amount of work. Mainly, we would have to come to grips somehow with the fact that the S_α are not all manifolds. We need some property possessed by S_0 such that:

(14) for sets with this property the operations A, B, C, and D make sense,

(15) the property is preserved by A, B, C, and D, and

(16) the number c, defined by (12), increases as the operations are applied.

§Intro. D.3.a. "Relative Semialgebraicity"

The property to use is a sort of "relative semi-algebraicity". Namely, each set S_α which we will encounter will be of the following kind: There will exist a corresponding set X_α contained in some multijet manifold ${}_{r_\alpha}J^{k_\alpha}(I\times P,\ I\times N)$ (see Ch.I, §D for the definitions) such that

$$S_\alpha = \left\{ (x,F) \in (I\times P)^{(r_\alpha)} \times C \;\middle|\; {}_{r_\alpha}j^{k_\alpha}(F)(x) \in X_\alpha \right\}.$$

The set X_α wil be "closed semi-algebraic" relative to P and N in the following sense: that in terms of (local coordinates in ${}_{r_\alpha}J^{k_\alpha}(I\times P,\ I\times N)$ determined by) any local coordinates in P and N, X_α is defined by polynomial equations and non-strict ($\geq$) inequalities. Note that ${}_rJ^k(I\times P,\ I\times N)$ is not itself an algebraic or semi-algebraic variety; this is the reason for the quotation marks above and the word "relative".

In verifying (14), (15), and (16) for this property, we can now shift our attention to the universal examples — the sets in ${}_{r_\alpha}J^{k_\alpha}(\mathbb{R}\times\mathbb{R}^p, \mathbb{R}\times\mathbb{R}^n)$ which are defined by these same equations and inequalities. (In practice, in Chapter II, we will do things a little differently. In order to avoid semi-algebraic sets we will forget the inequalities — this turns out not to matter — and consider the complex algebraic varieties of complex multijets from $\mathbb{C}\times\mathbb{C}^p$ to $\mathbb{C}\times\mathbb{C}^n$ defined by the equations.)

Now, for Operation A (14), (15), and (16) are pretty clear. For B (14) and (15) are clear, while (16) turns out to be a consequence of the hypothesis $n-p\geq 3$. (Recall the computation of $\text{codim}(p\pi_j S', P\times C)$). For C, (14) depends on the notion of (Zariski) tangent space of a (possibly singular) algebraic variety; (15) is then pretty obvious and (16) again uses that $n-p\geq 3$. (The proof of (16) for C is a bit of a nuisance — see Lemma 77, §II.C.2 below.)

§Intro. D.3.b. The Hilbert variety.

The fact that (15) holds for Operation D is perhaps the most interesting feature in all of this, and we will now take a few pages to give an idea of why this is so. The key is the "Hilbert variety", an idea borrowed from algebraic geometry.

In a C^∞ manifold M (such as $I\times P$) consider the set $\mathrm{Hilb}^r(M)$ of all "zero-dimensional subvarieties of length r" — that is, ideals I in the ring $C^\infty(M)$ of functions such that the vector-space dimension of $C^\infty(M)/I$ is r. (Thus the obvious points in $\mathrm{Hilb}^r(M)$, corresponding to the "reduced subvarieties" — sets of r distinct points — are the ideals $I = \bigcap_{i=1}^{r} m_{x_i}$, where $(x_1,\cdots x_r) \in M^{(r)}$ and m_x is the ideal $\{f \in C^\infty(M) \mid f(x) = 0\}$.) $\mathrm{Hilb}^r(M)$ can be given some sort of singular C^∞ structure. In fact it is "semi-algebraic" relative to M in much the same sense in which the sets X_α are. $\mathrm{Hilb}^r(M)$ admits a proper map "supp" to the symmetric product $\mathrm{Sym}^r(M)$, which is another "semi-algebraic space relative to M". (For any ideal $I \in \mathrm{Hilb}^r(M)$, $\mathrm{supp}(I)$ depends on the primary decomposition of I : If $I = \bigcap_i q_i$ for some collection of primary ideals $q_i \subset m_{x_i}$, then the unordered r-tuple $\mathrm{supp}(I)$ contains x_i with multiplicity $\mathrm{codim}(q_i, C^\infty(M))$.)

There is an obvious tautological vector bundle E_r of rank r over $\mathrm{Hilb}^r(M)$; its fiber over $I \in \mathrm{Hilb}^r(M)$ is $C^\infty(M)/I$. (At least this is obvious in the category of sets; we have said nothing, and will say nothing, about how the topologies and other structures on these sets are defined.) Every function $f \in C^\infty(M)$ induces an obvious section $\sigma(f)$ of E_r.

REMARK 3. It is clear that this construction has much to do with the multijet sets ${}_rJ^m(M,\mathbb{R})$ and the sections ${}_rj^m(f)\colon M^{(r)} \longrightarrow {}_rJ^m(M,\mathbb{R})$. It is good to have both the Hilbert-variety point of view and the multijet point of view; the former has the disadvantage of introducing singular spaces, while it has the advantage of handling all of M^r at once and not just $M^{(r)}$.

Let us try to understand Operation D from this viewpoint. For simplicity we will think of functions $M \to \mathbb{R}$ instead of maps $I\times P \to I\times N$. We will also restrict our attention to the extreme case of totally diagonal limit points $(x,x,\cdots x) \in M^r$.

Suppose $X \subset {}_rJ^q(M,\mathbb{R})$ is a set of multijets which is "closed semi-algebraic" with respect to local coordinates on M. Thus X determines a set

$$S_X = \left\{ (x,f) \in M^{(r)} \times C^\infty(M) \,\middle|\, {}_rj^m(f)(x) \in X \right\}.$$

In analogy with the definition of Operation D, set

$$D(S_X) = \left\{ (x,f) \in M\times C^\infty(M) \,\middle|\, (\underbrace{x,x,\cdots x}_{r\ \text{times}},f) \in \overline{S}_X \right\},$$

where "bar" denotes closure in $M^r \times C^\infty(M)$.

We wish to find a "closed semi-algebraic" set $X' \subset J^{q'}(M,\mathbb{R})$ for some q', such that

$$D(S_X) = S_{X'} \quad .$$

Let $q' = r\binom{q+m}{m} - 1$ where $m = \dim M$. The equation

$$\Phi(x_1,\cdots x_r) = \bigcap_{i=1}^{r} m_{x_i}^{q+1}$$

defines a map (algebraic with respect to coordinates in M)

$$M^{(r)} \xrightarrow{\Phi} \mathrm{Hilb}^{q'+1}(M) \quad .$$

The vector bundle ${}_rJ^q(M,\mathbf{R})$ over $M^{(r)}$ can be identified with $\Phi^*E_{q'+1}$ in an obvious way. Φ maps $M^{(r)}$ properly to the "open semi-algebraic" set

$$\mathcal{U} \underset{\mathrm{def}}{=} \left\{ I \in \mathrm{Hilb}^{q'+1}(M) \;\middle|\; \text{in supp } (I) \text{ no point has multiplicity } > \binom{q+m}{m} \right\} \quad .$$

Thus Φ_*X is "closed semi-algebraic" in $E_{q'+1}\big|_{\mathcal{U}}$, and the closure $\overline{\Phi_*X}$ in $E_{q'+1}$ is "closed semi-algebraic" in $E_{q'+1}$.

Let $\Delta \subset \mathrm{Hilb}^{q'+1}(M)$ be $\mathrm{supp}^{-1}(\Delta_M)$, where $\Delta_M \subset \mathrm{Sym}^{q'+1}M$ is the (strict) diagonal. Let s be the map which makes the diagram

$$\begin{array}{ccc} \Delta & \xrightarrow{s} & M \\ \downarrow & & \downarrow \cong \\ & & \Delta_M \\ \downarrow & & \downarrow \\ \mathrm{Hilb}^{q'+1}M & \xrightarrow{\mathrm{supp}} & \mathrm{Sym}^{q'+1}M \end{array}$$

commute. The vector bundle $E_{q'+1}\big|_{\Delta}$ can be identified with a quotient bundle of $s^*J^{q'}(M,\mathbb{R})$, because any ideal of codimension $q'+1$ supported at a point x contains $m_x^{q'+1}$. Let

$$s^*J^{q'}(M,\mathbb{R}) \xrightarrow{Q} E_{q'+1}\big|_{\Delta}$$

be the quotient map. Set

$$X' = s_*Q^{-1}\left(\overline{\Phi_*X} \cap E_{q'+1}\big|_{\Delta}\right),$$

a "closed semi-algebraic set" in $J^{q'}(M, \mathbb{R})$.

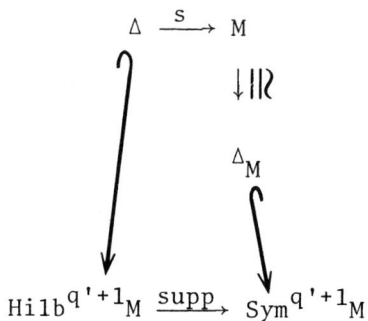

We claim that $S_{X'} = D(S_X)$. First suppose $(x,f) \in D(S_X)$. Thus

$$(x,\dots,x,f) = \lim_{\nu=1}^{\infty} (x^{\nu},f^{\nu}) , \quad \text{where}$$

$$x^{\nu} = (x_1^{\nu},\dots x_r^{\nu}) \in M^{(r)}$$

$$f^{\nu} \in C^{\infty}(M)$$

$${}_r j^m(f^{\nu})(x^{\nu}) \in X .$$

By the properness of

$$\mathrm{Hilb}^{q'+1}M \xrightarrow{\text{supp}} \mathrm{Sym}^{q'+1}M$$

the sequence Φx^{ν} has at least one limit point I in $\mathrm{Hilb}^{q'+1}M$. We have

$$\sigma(f^{\nu})(\Phi x^{\nu}) \in \Phi_* X, \quad \text{so that in the limit}$$

$$\sigma(f)(I) \in \overline{\Phi_* X} \cap E_{q'+1}\Big|_{\Delta} \quad \text{and}$$

$$j^{q'}(f)(x) \in X' , \quad \text{i.e.}$$

$$(x,f) \in S_{X'} \quad \text{as claimed.}$$

Now suppose $(x,f) \in S_{X'}$, i.e.

$$j^{q'}(f)(x) \in X' .$$

This means that for some $I \in s^{-1}(x)$ we have

$$\sigma(f)(I) \in \overline{\Phi_* X} \ .$$

Thus for some sequences $x^\nu = (x_1{}^\nu, \cdots x_r{}^\nu) \in M^{(r)}$ and $f^\nu \in C^\infty(M)$ we have

$$\begin{aligned} &\sigma(f^\nu)(\Phi x^\nu) \in \Phi_* X \text{ and} \\ &\lim \sigma(f^\nu)(\Phi x^\nu) = \sigma(f)(I). \end{aligned} \tag{17}$$

In fact the f^ν can be chosen to converge to f. Thus (17), which is equivalent to $(x^\nu, f^\nu) \in S_X$, implies $(x,f) \in D(S_X)$.

For mappings from M into a real vector space rather than into $\mathbb{R}$ a similar description of $D(S_X)$ applies, and it can be extended by using local coordinates to apply to mappings from one manifold to another, for example $I \times P$ to $I \times N$. Also, if the "semi-algebraicity" of X is only assumed to hold in those coordinate systems in $I \times P$ (resp. $I \times N$) for which the last p (resp. last n) coordinates are functions on P (resp. N), then the same will hold for X' in these same coordinate systems.

REMARK 4. Glaeser ([Gl]) has given a definition of the Hilbert variety $\mathrm{Hilb}^q(M)$ and the vector bundle E_q, and an account of their most important properties. However, [Gl] is not directly applicable here because Glaeser considers $\mathrm{Hilb}^q(M)$ (in his notation, $\tilde{M}(q)$) merely as a topological space. In carrying out the program being sketched here it would be necessary to make a more detailed study of $\mathrm{Hilb}^q(M)$ and E_m and of their relationship with symmetric products and multijet manifolds, keeping track of "relative semialgebraicity" throughout. In this thesis we use the down-to-earth results of §I.E and §II.B.4 instead of this hypothetical machinery.

§Intro. D.4. Keeping track of the ordering.

Now consider the collection $\{S_\alpha\}$ obtained from S_0 by Operations A,B,C, and D and consider the sets $W_\alpha \subset S_\alpha$ defined by (13). Notice that for each α, i, and j $(i,j \in \{1,\cdots r_\alpha\})$ the statement

$$p\pi_i x = p\pi_j x \tag{18}$$

holds for one $x \in W_\alpha$ if and only if it holds for all $x \in W_\alpha$. To see this, set

$$S_\beta = B_{id,id,i,j}(S_\alpha, S_\alpha) = \left\{ x \in S_\alpha \,\middle|\, p\pi_i x = p\pi_j x \right\}$$

Thus either (Case 1) $S_\beta \subsetneq S_\alpha$ or (Case 2) $S_\beta = S_\alpha$. In Case 1 $W_\alpha \cap S_\beta = \phi$ by (13), so that (18) fails for all $x \in W_\alpha$. In Case 2 (18) holds for all $x \in S_\alpha$ and a fortiori for all $x \in W_\alpha$. Write $i \sim j$ if (18) holds for $x \in W_\alpha$. For each $x \in W_\alpha$ the point $p\pi_i x \in P \times C$ has several points $\pi_j x \in I \times P \times C$ "stacked above it" corresponding to the several j such that $j \sim i$.

Likewise, working with N instead of P, if we define

$$(I \times P)^{(r_\alpha)} \times C \xrightarrow{\pi_i^!} (I \times N) \times C$$

$$I \times N \times C \xrightarrow{p'} N \times C$$

by

$$\pi_i^!(x_1, \cdots x_{r_\alpha}, F) = (F(x_i), F)$$

$$p'(t, x, F) = (x, F) ,$$

then the statement

$$(19) \qquad p\pi_i^! x = p\pi_j^! x$$

will hold for all $x \in W_\alpha$ if it holds for one. For if we set

$$S_\beta = B_{id,\phi,1,i}(S_\alpha, S_o)$$

$$\phi(1) = i \qquad \phi(2) = j$$

then we can argue with two cases much as we did above. Define $i \simeq j \Longleftrightarrow$ (19) holds for $x \in W_\alpha$. Then for $x \in W_\alpha$ there are points $\pi_j' x \in I \times N \times C$ for all j such that $j \simeq i$, stacked over $p' \pi_i' x$ in some order.

It will be convenient to decompose W_α further into sets $W_\alpha^{(D,R)}$ so as to keep track of the order of the stacking. Suppose that D (for "domain") is a binary relation on $\{1, \cdots r_\alpha\}$ such that within each $\sim$-class it is a strict total ordering and points in distinct $\sim$-classes are unrelated, and that R (for "range") is another such for the $\simeq$-classes. Set

$$W_\alpha^{(D,R)} = \left\{ x \in W_\alpha \,\middle|\, \forall_{i,j} \ \pi_i x \text{ is below } \pi_j x \text{ if } iDj \text{ and } \pi_i' x \text{ is below } \pi_j' x \text{ if } iRj \right\}.$$

Then we have decomposed each W_α into finitely many open pieces. Their images $p\pi_i W_\alpha^{(D,R)}$ are the strata which we finally use.

§Intro.D.5. Using the stratification.

We will now hint at how the stratification can be used to prove Theorem B. The key is the inductive step mentioned in (10). In the course of very briefly describing that step we will explain the distinction between "good" and "bad" strata, thus shedding light on (7) and (8).

Let $p\pi_i W_\alpha^{(D,R)}$ be some stratum, and assume that $(9)_{\Sigma'}$ holds for all smaller strata Σ' (i.e., for all $\Sigma' = p\pi_{i'} W_{\alpha'}^{(D',R')}$ such that $c(S_{\alpha'})$ is greater than $c(S_\alpha)$). Choose $i_o \in \{1,\cdots r_\alpha\}$ such that i_o is maximal with respect to D in its $\sim$-class and maximal with respect to R in its $\simeq$-class. This is always possible unless there exist distinct elements

$$i_1, i_2, \cdots i_{2a} \in \{1,\cdots r_\alpha\} \quad \text{for some} \quad a \geq 1$$

such that

(20) $$i_1 \sim i_2 \simeq i_3 \sim \cdots \simeq i_{2a-1} \sim i_{2a} \simeq i_1 .$$

The nonexistence of such a "closed loop" is one half of the definition of "goodness" of S_α (or of its strata $p\pi_i W_\alpha^{(D,R)}$). In fact the largest bad S_α is

$$B_{id,id,1,2}(S_o,S_o) = \{(x_1,x_2,F) \mid Fx_1 \text{ is below } Fx_2 \text{ and } px_1 = px_2\} ,$$

for which $1 \sim 2 \simeq 1$. In this example $c=n-p-2$, so the strata have codimension $n-2$, in agreement with (8).

Now suppose that S_α is good (we have not yet finished saying what this means) and that i_o has been chosen as above. Consider the stratum $\Sigma = p\pi_{i_o} W_\alpha^{(D,R)}$. Let $\tilde{\Sigma}$ be its "uppermost lifting" : $\tilde{\Sigma} = \pi_{i_o} W_\alpha^{(D,R)}$. Let $\hat{\tilde{\Sigma}}$ be the set of points "above" $\tilde{\Sigma}$ in $I \times P \times C$:

$$\hat{\tilde{\Sigma}} \underset{\text{def}}{=} \left\{ (t,x,F) \in I \times P \times C \;\middle|\; \exists_{t'<t} \; (t',x,y) \in \tilde{\Sigma} \right\}$$

The set $\hat{\tilde{\Sigma}}$ is disjoint from $\pi_j S_\beta$ for all β and j. Indeed, if $(t,x,F) \in \pi_j S_\beta \cap \tilde{\Sigma}$ then for some r and some ϕ and ϕ'

$$\{1,\cdots r_\alpha\} \xrightarrow{\phi} \{1,\cdots r\}$$

$$\{1,\cdots r_\beta\} \xrightarrow{\phi'} \{1,\cdots r\}$$

we must have

$$(t,x,F) \in \pi_{\phi(j)} S_\gamma \quad \text{where}$$

$$S_\gamma = B_{\phi,\phi',i_o,j}(S_\alpha,S_\beta) \; .$$

A familiar kind of argument applies: Case 1: $S_\gamma = S_\alpha$ up to a permutation of $\{1,\cdots r\}$. Case 2: S_γ is smaller than S_α. Case 2 is ruled out by definition of W_α. In Case 1 we have that for some $i \in \{1,\cdots r_\alpha\}$

$$(t,x,F) = \pi_i \bar{x}\ , \quad \text{where}$$

$$(t',x,F) = \pi_{i_o} \bar{x}\ , \ t'<t,\ \bar{x} \in W_\alpha^{(D,R)}\ .$$

Thus $i_o D i$, contradicting the choice of i_o. In particular $\hat{\Sigma}$ is disjoint from $\pi_1 S_0$.

Also, $\tilde{\Sigma}$ is itself disjoint from $\pi_1 S_0$ by the same sort of argument, using R instead of D. (Suppose $x \in \pi_{i_o} W_\alpha^{(D,R)} \cap \pi_1 S_0$. Then $x = \pi_{i_o} \phi^* \bar{x}$ for some $\bar{x} \in S_\beta$ where for suitable ϕ and ϕ'

$$S_\beta = B_{\phi,\phi',i_o,1}(S_\alpha, S_o)\ .$$

Case 1: $S_\beta = S_\alpha$ up to permutation. Case 2: S_β is smaller. Case 2 contradicts the definition of W_α. Case 1 implies $i_o R i$ for some i, contradicting the choice of i_o.)

Now the fact that the union $\hat{\hat{\Sigma}} \cup \tilde{\tilde{\Sigma}}$ is disjoint from $\pi_1 S_0$ means that its image in $I \times N \times C$:

$$T \underset{\text{def}}{=} \left\{ (Fx, F) \in (I \times N) \times C \,\middle|\, (x, F) \in \hat{\hat{\Sigma}} \cup \tilde{\tilde{\Sigma}} \right\}$$

has no points above it in the full image; i.e. the set

$$\hat{T} \underset{\text{def}}{=} \left\{ (t, x, F) \in I \times N \times C \,\middle|\, \exists_{t' < t} (t', x, F) \in T \right\}$$

is disjoint from

$$\left\{ (Fx, F) \in (I \times N) \times C \,\middle|\, (x, F) \in (I \times P) \times C \right\}.$$

In pictures, representing C as a point and representing P and N by:

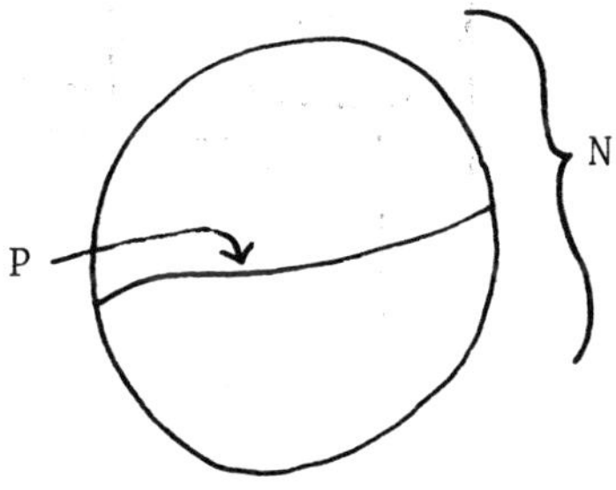

so that one concordance F looks like

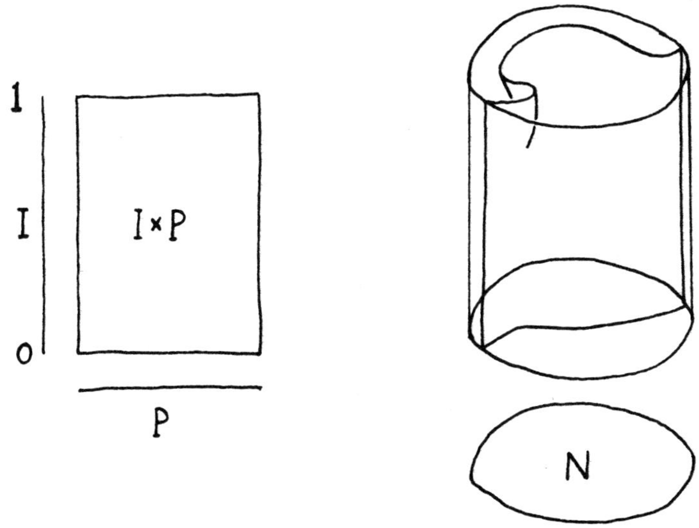

we have something like

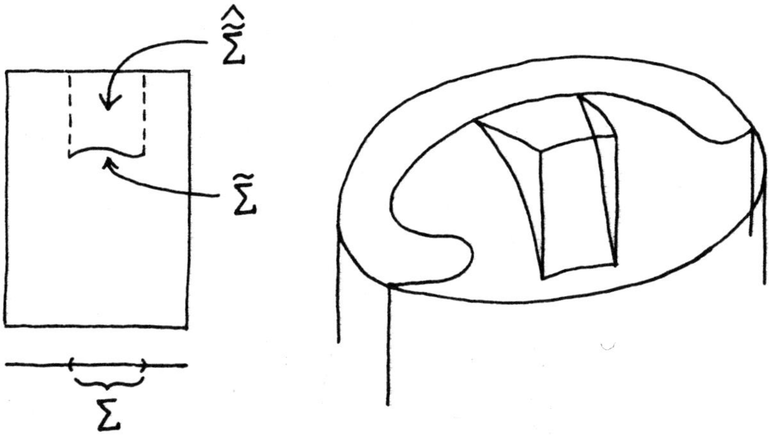

On the RHS the bottom face of the wedge-shaped figure is $T = F(\hat{\Sigma} \cup \tilde{\Sigma})$, and the solid wedge $(\hat{T} \cup T)$ is supposed to be otherwise disjoint from $F(I\times P)$.

Now pull all of this back by the given map $y \longmapsto F_y$ from D^s to C. Thus, for example, the set $W_\alpha^{(D,R)} \subset (I\times P)^{(r_\alpha)} \times C$ yields a set in $(I\times P)^{(r_\alpha)} \times D^s$, and if suitable general position conditions are met then the stratification of $P\times C$ yields a stratification of $P\times D^s$. Let us denote the pulled back objects by the same names as the originals. Thus the pictures above are pictures of subsets of $I\times N\times D^s$ and $I\times P\times D^s$. (We have drawn the case $s=0$.)

We have to describe the fibered isotopy referred to in (10). It is suggested by the following pictures:

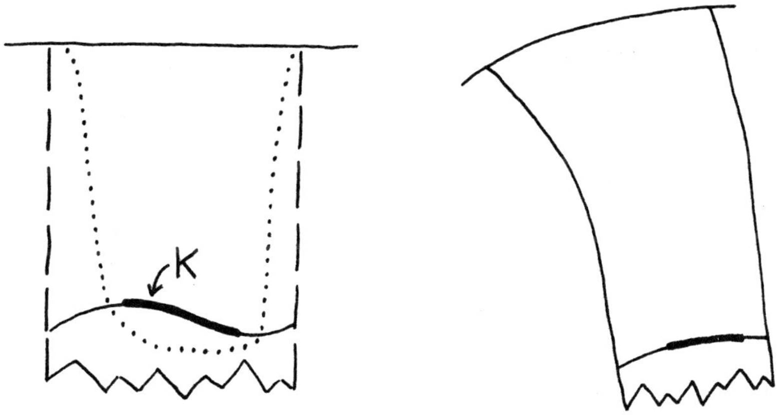

(a.) before

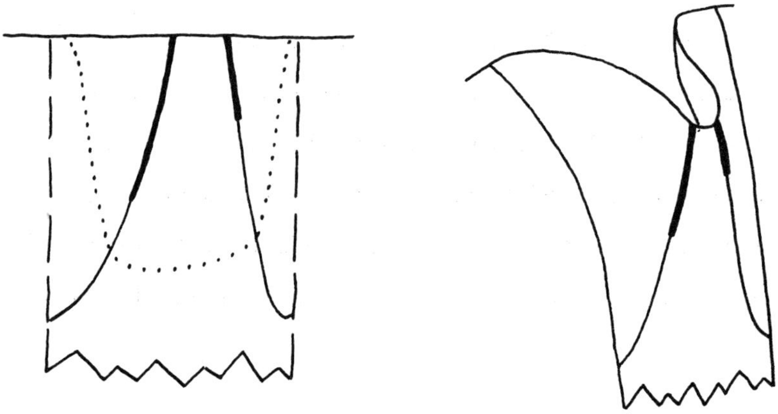

(b.) during

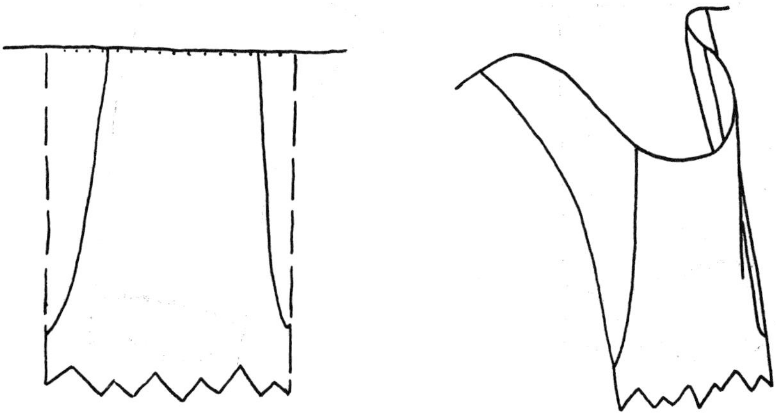

(c.) after

The LHS of (a) is a closeup view of a neighborhood of $\hat{\tilde{\Sigma}} \subset I\times P\times D^s$. In Σ something unpleasant is taking place in a compact set K. We want to eliminate the unpleasantness by "eliminating K". We start by choosing a function $\phi\colon P\times D^s \to I$ whose graph (the dotted line) goes below K, but just barely. Next we choose a family Φ^u $(o\le u\le 1)$ of embeddings of $I\times P\times D^s$ into itself such that

> Φ^o = identity,
>
> every point in $I\times P\times D^s$ which moves is above or near the graph of ϕ,
>
> every point in $I\times P\times D^s$ moves vertically downward if it moves at all, and
>
> the final image $\Phi^1(I\times P\times D^s)$ has as its "upper boundary" $\Phi^1(1\times P\times D^s)$ = graph of ϕ.

The LHS of (b) (resp (c)) shows the preimages of $\tilde{\Sigma}$, $\hat{\tilde{\Sigma}}$, K, and the graph of ϕ under $\Phi^{1/2}$ (resp. Φ^1).

Consider the family $F\circ\Phi^u$ of embeddings of $I\times P\times D^s$ in $I\times N\times D^s$; it is not a fibered isotopy of concordances because $F(\Phi^u(1\times P\times D^s)) \not\subset 1\times N\times D^s$. As u goes from 0 to 1 the image $F\Phi^u(I\times P\times D^s)$ shrinks; its top edge $F\Phi^u(1\times P\times D^s)$ sweeps through a small neighborhood (in $F(I\times P\times D^s)$) of T.

We would like the image of $1\times P\times D^s$ to stay inside of $1\times N\times D^s$. Now, $\hat{T}$ is just the set of points directly above T, i.e. between T and $1\times N\times D^s$, so the fact that $\hat{T} \cap F(I\times P\times D^s) = \emptyset$ suggests that there is room to maneuver somehow and make $F\circ\Phi^u$ into a fibered isotopy of concordances.

In fact, if we have been careful enough in choosing Φ^u, there is : it is possible to make a family Ψ^u of embeddings of $I\times N\times D^s$ into itself such that

> Ψ^o = identity,
> points which move are near $\bar{\hat{T}}$, and
> every point which moves moves vertically downward;

and such that Ψ^u and Φ^u are compatible in the sense that:

> for each u the image $\Psi^u(1\times N\times D^s)$ cuts $F(I\times P\times D^s)$ transversely in the set $F\Phi^u(1\times P\times D^s)$.

Thus there is a fibered isotopy of concordances F^u such that

$$F\Phi^u = \Psi^u F^u.$$

The RHS's of (a), (b) and (c) show closeups of the relevant parts of the images of $F^0, F^{\frac{1}{2}}$, and F^1.

The set K is the set of points $p\pi_{i_0} x \in \tilde{\Sigma}$ such that for some i and $\Sigma_i = p\pi_i W_\alpha^{(D,R)}$ $(9)_{\Sigma_i}$ fails at $p\pi_i x$. Write $K=K(F)$. Because of the way in which the stratification was defined and the fact that F^u is made up of a vertical motion (Φ^u) of the domain and a vertical motion (Ψ^u) of the range, it follows that $K(F^u) = (\Phi^u)^{-1}(K(F))$. Thus "eliminating K" (i.e. making $(\Phi^1)^{-1}(K)$ empty) is indeed the same as "eliminating the unpleasantness" (i.e. making $K(F^1)$ empty). The fact that K is compact comes from the inductive hypothesis ($(9)_{\Sigma'}$ for lower strata Σ').

What subtlety there is in this argument lies in the choice of ϕ and Φ^u. Some care is required here for two reasons.

The first is that it is only T, and not a neighborhood of T in $F(I\times P\times D^s)$, which satisfies $\hat{T} \cap F(I\times P\times D^s) = \emptyset$. This means paying close attention to the lower strata.

The second is that in order to make the argument work we need also an "infinitesimal analogue" of the statement "$\hat{\tilde{\Sigma}}$ is disjoint from $\pi_1 S_0$". Namely, if we are in the case where

$$\ker D(pF)_{x_{i_0}} \neq \{0\}$$

for $(x_1,\cdots x_r,F) \in W_\alpha$ (see §D.2.e) then we need the vector field η_{α,i_0} along Σ to be nowhere tangent to Σ. This finally explains what Operation C is good for: C implies that the only way in which the condition above can fail is for η_{α,i_o} to be zero identically, i.e. for ξ_{α,i_o} to be vertical identically, i.e. for $D(p\circ F)_{x_{i_o}} \cdot \frac{\partial}{\partial t}$ to be zero identically for $(x_1,\cdots x_r,F) \in S_\alpha$. If this is so then S_α will be called "bad". The final definition of "bad" is then:

> A set S_α is <u>bad</u> if either there is a closed loop as in (20) or for some i $(1\le i\le r_\alpha)$ we have $D(p\cdot F)_{x_i} \cdot \frac{\partial}{\partial t} = 0$ for all $(x_1,\cdots x_r,F) \in S_\alpha$.

The "largest" S_α which is bad in the second way is

$$S_\alpha = \left\{ (x,F) \in (I\times P) \times C \mid D(p\circ F)\cdot\frac{\partial}{\partial t} = 0 \right\},$$

which has $c(S_\alpha) = n-p-1$, so that its strata have codimension $n-1 \ge n-2$, as required by (8).

The arguments hinted at above are found in §I.D. and §III.C. In I.D. we show how to make ψ^u and hence F^u if ϕ^u is given and has suitable properties. (Actually we set

$$\Phi^u(t,x,y) = ((1-u)t + u\,\phi^u(x,y),x,y)$$

$$\Psi^u(t,z,y) = ((1-u)t + u\,\psi^u(z,y),z,y)$$

and speak of constructing ψ^u if ϕ^u has certain properties. Any ϕ^u with those properties is called a <u>sunny collapse</u>. See Def 28, §I.D.)

In §III.C, which constitutes the proof of what corresponds to (10), we construct a sunny collapse (in §§3,4, and 5) to do the job at hand. (There are two other sunny collapses constructed in Chapter III — a preliminary one in §III.C.2 and a final one in §III.D to "make $(9)_\Sigma$ true for the codimension-zero stratum Σ" — but the crucial one is the one we have been describing.

Chapter I. Preliminaries

§I.A. Definitions, etc.

§I.A.1. Manifolds.

In this section we collect a few definitions, conventions, notations, and simple facts.

We use the letter "p" with subscripts to denote the projection of a product onto its factors, e.g.

$$X\times Y\times Z \xrightarrow{p_1} X$$

$$X\times Y\times Z \xrightarrow{p_{2,3}} Y\times Z .$$

Except when the contrary is stated, all manifolds, maps between manifolds, functions, embeddings, and immersions will be smooth (= C^∞). Manifolds will be allowed to have nonempty boundary, or even corners.

The differential of a smooth real-valued function f is written "df"; the derivative of a map F between manifolds is written "DF". In each case evaluation on a vector is written with a dot:

$$(df)\cdot\xi \quad \text{is a number}$$

$$(DF)\cdot\xi \quad \text{is a vector}$$

We use the same notation in the complex case for holomorphic functions and maps and vectors in the complex tangent bundle.

An embedding of manifolds $F : X \longrightarrow Y$ is <u>proper</u> if F is transverse ("$\pitchfork$") to the k-codimensional corner set of Y and if the preimage of this set is the k-codimensional corner set of X for all $k>0$. (If X and Y have no corners then this just means $F \pitchfork \partial Y$ and $F^{-1}(\partial Y) = \partial X$.) A submanifold is proper if it is the image of a proper embedding.

A <u>fibered embedding</u> of X in Y over Z (X, Y, and Z manifolds) is an embedding $F\colon X \times Z \longrightarrow Y \times Z$ making the triangle commute

$$\begin{array}{ccc} X\times Z & \xrightarrow{\ F\ } & Y\times Z \\ & {\scriptstyle p_2}\searrow \quad \swarrow{\scriptstyle p_2} & \\ & Z & \end{array}$$

If we write $F(x,z) = (F^z(x), z)$, then each F^z

$$X \xrightarrow{\ F^z\ } Y \qquad z \in Z$$

is an embedding. In the special case $Z = I$ $(= [0,1])$ we call a family

$$X \xrightarrow{\ F^u\ } Y \qquad 0 \le u \le 1$$

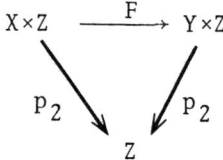

depending smoothly on u an <u>isotopy</u> if $(x,u) \longmapsto (F^u(x),u)$ is an embedding, or equivalently if each F^u is an embedding. We will even have occasion to mix these two kinds of terminology and speak of a <u>fibered isotopy</u> of embeddings of X in Y over Z:

$$X\times Z \xrightarrow{F^u} Y\times Z \qquad 0\le u\le 1$$

(This just means that $(x,z,u) \longmapsto (F^u(x,z),u)$ is a fibered embedding of X in Y over $Z\times I$.)

All of the above can be repeated for diffeomorphisms. Thus we speak of <u>fibered diffeomorphisms</u> of X over Z, etc. An isotopy of diffeomorphisms of X

$$X \xrightarrow{F^u} X \qquad 0\le u\le 1$$

such that F^0 = identity is, as usual, simply called an isotopy of X.

Now let N be a manifold and $P \subset N$ a proper submanifold. A <u>concordance</u> of P in N is a proper embedding

$$I\times P \xrightarrow{F} I\times N$$

such that

$$F\big|_{0\times P \cup I\times\partial P} = \text{inclusion, and}$$

$$F^{-1}(1\times N) = 1\times P$$

A <u>fibered concordance</u> of P in N over Z is a fibered embedding F of $I\times P$ in $I\times N$ over Z such that for each $z \in Z$ the map

$$I\times P \xrightarrow{F^z} I\times N$$

is a concordance. A <u>fibered isotopy</u> of <u>concordances</u> of P in N over Z is a fibered isotopy F^u of embeddings of $I\times P$ in $I\times N$ such that for each u $(0\leq u\leq 1)$ F^u is a fibered concordance of P in N over Z.

The space of all concordances of P in N is denoted $C(P,N)$. On $C(P,N)$ and any other space of C^∞ mappings the topology is the (strong) Whitney topology (see [Hi] Chapter 2).

§I.A.2. Algebraic Varieties.

Here we collect a few definitions and elementary facts from complex algebraic geometry. Most of what we will use is more or less contained in Chapter I of [S].

An algebraic variety for us is a quasi-projective variety over $\mathbb{C}$, i.e. a subset of some complex projective space $\mathbb{P}^N$ defined by homogeneous polynomial equations and inequations. A projective variety is defined by equations only. The Zariski topology in $\mathbb{P}^N$ has for closed sets the projective varieties. The complex topology is the usual topology of $\mathbb{P}^N$ as a complex manifold. The Zariski and complex topologies on an arbitrary algebraic variety in $\mathbb{P}^N$ are by definition the corresponding subspace topologies.

FACT 5: (Noetherian property) In an algebraic variety X every infinite descending chain of Zariski closed subsets $X_1 \supset X_2 \supset X_3 \supset \cdots$ must terminate ($X_n = X_{n+1}$ for all large n). (See [Har] Exercise 2.5(a), p.11.)

FACT 6: If $f:X \to Y$ is a morphism of algebraic varieties and $\overline{f(X)}$ is the Zariski closure of the image in Y, then some Zariski dense open subset of $\overline{f(X)}$ is contained in $f(X)$. (See [S], Thm. 6, page 60.)

FACT 7: In the same situation, $\overline{f(X)}$ equals the complex closure of $f(X)$ in Y. This follows from Fact 6.

REMARK 8. The <u>tangent space</u> T_xX of an algebraic variety X at a point $x \in X$ is the Zariski tangent space (see [S] p.75). The union $\bigcup_{x \in X} T_xX = TX$ has a structure of algebraic variety and a morphism $\pi\colon TX \to X$ such that $\pi^{-1}x = T_xX$. The <u>singular set</u> of a d-dimensional algebraic variety X will by definition consist of the points x where $\dim(T_xX) > d$ <u>and</u> the points which lie on components of X which have dimension $< d$. The singular set is Zariski closed and has dimension $< d$. Its complement in X (the set of <u>non-singular</u> points) is a nonempty d-dimensional complex manifold. (For all of this see [S], pp. 75-78.) The singular set is in fact defined for any complex analytic variety, and is invariant under local complex diffeomorphisms of the ambient space.

The next two facts have to do with subsets $S \subset E$ of algebraic vector bundles $E \xrightarrow{\pi} X$.

FACT 9: If $S \subset E$ is Zariski closed and for each $x \in X$ $\pi^{-1}(x) \cap S$ is a vector subspace of $\pi^{-1}(x)$, then $\dim(\pi^{-1}(x) \cap S)$ achieves its minimum on a Zariski open set in X. (See [S] Thm. 7, p.60.)

FACT 10: If in addition $\dim(\pi^{-1}(x) \cap S)$ is a constant k and X is nonsingular, then S is a vector subbundle of E. (This is a local fact, so w.l.o.g. $E = \mathbb{C}^n \times X$, $\pi : E \to X$ the projection. Let $\mathbb{G}$ be the Grassmannian of k-planes in $\mathbb{C}^n$. Set

$$\Gamma = \{(W,x) \in \mathbb{G}\times X \mid W\times\{x\} \subset S\}.$$

Clearly Γ is Zariski closed in $\mathbb{G}\times X$. The projection $\Gamma \to X$ is a bijection. Since X is normal (in fact, nonsingular) the inverse $X \to \Gamma$ is a morphism. This implies that S is a subbundle.)

Finally, we need:

FACT 11: If a morphism $f\colon X \to Y$ between nonsingular varieties is nowhere an immersion (i.e., $\ker((DF)(x)) \neq 0 \quad \forall_{x \in X}$) then every nonempty fiber of f has positive dimension. (W.l.o.g. X is irreducible. By Sard's Theorem ([Hi], p.69)

$f(X)$ cannot contain a complex manifold of dimension $\dim(X)$. Fact 6 then implies $\dim \overline{f(X)} < \dim(X)$. The result now follows from [S] Thm. 7, p.60, without much trouble: Apply that theorem to the projection

$$\overline{\{(x, f(x)) \mid x\in\overline{X}\}} \subset \overline{X}\times\overline{f(X)} \longrightarrow \overline{f(X)}.)$$

§I.B. <u>(a+1)-ads and Connectivity.</u>

In this section we review the notion of "(a+1)-ad" (more precisely, (a+1)-ad of topological spaces). We introduce the notion of "$\underline{k}$-connectedness" of an (a+1)-ad, where $\underline{k}=\{k_T\}$ is a collection of integers indexed by the subsets T of $\{1,\cdots a\}$. Using this notion we state a very general form of "homotopy excision" (Lemma 18) and deduce it quite formally from a result of Barratt and J.H.C. Whitehead (Lemma 19 – homotopy excision for "pushout (a+1)-ads").

Lemma 18 will be used in the proof of Theorem D to reduce a statement about homotopy groups of an (a+1)-ad of concordance spaces to a statement about homotopy groups of certain pairs.

A slightly different proof of Lemma 18 appears in [Go1] (Corollary 3.2).

§I.B.1. (a+1)-ads.

DEF 12. Let $a \geq 0$ be an integer. A topological (a+1)-ad $\underline{X}$ consists of a topological space X and subspaces $X_1, \cdots X_a$ of X. We write

$$\underline{X} = (X; X_1, \cdots X_a) = \left(X; \{X_j\}_{j=1}^{a}\right)$$

$\underline{X}$ is a CW (a+1)-ad if X is CW and each X_j is a subcomplex. $\underline{X}$ is an open (a+1)-ad if each X_j is open in X. If $\underline{X}$ is equipped with a basepoint in $\bigcap_{j=1}^{a} X_j$ then $\underline{X}$ is a pointed (a+1)-ad.

In thinking of an (a+1)-ad $(X; \{X_j\}_{j=1}^{a})$ it is good to keep in mind all of the 2^a multiple intersections

$$X_S \overset{\text{def}}{=} \bigcap_{j \in S} X_j \quad (S \subset \{1, \cdots a\}).$$

These spaces form a diagram in the shape of an a-dimensional cube. The "faces" of the cube are the (a-1)-dimensional cubes associated with certain a-ads $D_j^{\epsilon}X$. Namely for $1 \leq j \leq a$ define

$$D_j^1\underline{X} = (X; X_1, \cdots X_{j-1}, X_{j+1}, \cdots X_a)$$

$$D_j^0\underline{X} = (X_j; X_1 \cap X_j, \cdots X_{j-1} \cap X_j, X_{j+1} \cap X_j, \cdots X_a \cap X_j).$$

After defining homotopy groups for pointed (a+1)-ads we will obtain a long exact sequence (21) which shows that $\underline{X}$ measures the difference between $D_j^1\underline{X}$ and $D_j^0\underline{X}$.

§I.B.2. The Homotopy groups of a pointed (a+1)-ad.

By definition a map

$$(X;\{X_j\}) \xrightarrow{f} (Y,\{Y_j\})$$

between (a+1)-ads is a map from X to Y such that $f(X_j) \subset Y_j$ for all j.

Let the faces of the cube I^a be denoted $F_j^\epsilon I^a$:

$$F_j^\epsilon I^a = \left\{(t_1,\cdots t_a) \in I^a \;\middle|\; t_j = \epsilon\right\},\quad \epsilon=0 \text{ or } 1,\ 1\le j\le a .$$

If $=(X;\{X_j\})$ is a pointed (a+1)-ad with basepoint $*$, define a pointed space $\Phi(\underline{X})$ by

$$\Phi(\underline{X}) = \left\{\text{maps of } (a+2)\text{-ads}\right.$$
$$\left.(I^a;F_1^1I^a,\cdots F_a^1I^a, \bigcup_{j=1}^a F_j^0I^a) \to (X;X_1,\cdots X_a,\{*\})\right\}.$$

(The topology is compact-open; the basepoint is the constant map.) The homotopy groups of $\underline{X}$ are defined by:

DEF 13. $\pi_s(\underline{X}) = \pi_{s-a}\Phi(\underline{X})$ for $s \geq a$, for any pointed $(a+1)$-ad $\underline{X}$. Thus $\pi_s(\underline{X})$ is a pointed set for $s=a$, a group for $s=a+1$, an abelian group for $s \geq a+2$. π_s is a functor.

The maps $\pi_s D_j^0 \underline{X} \longrightarrow \pi_s D_j^1 \underline{X}$ induced by inclusion (for any pointed $(a+1)$-ad $\underline{X}$ and $1 \leq j \leq a$) fit into a long exact sequence

$$\cdots \to \pi_{a+1}(\underline{X}) \to \pi_a(D_j^0\underline{X}) \longrightarrow \pi_a(D_j^1\underline{X}) \longrightarrow \pi_a(\underline{X}) \to \pi_{a-1}(D_j^0\underline{X}) \to \pi_{a-1}(D_j^1\underline{X}) \tag{21}$$

because $\Phi(\underline{X})$ is homeomorphic to the homotopy fiber of the inclusion map $\Phi(D_j^0\underline{X}) \hookrightarrow \Phi(D_j^1\underline{X})$.

REMARK 14. As a first application of (21), $\pi_s(\underline{X}) = 0$ for all $s \geq a$ if for some j $D_j\underline{X} = D_j^1\underline{X}$, i.e., $X_j = X$.

REMARK 15. Suppose $X_j \subset X_i$ for some $j \neq i$. Then since $X_j = X_j \cap X_i$ Remark 14 applies to $D_j^0\underline{X}$. Thus $\pi_s D_j^0 \underline{X} = 0$ for all $s \geq a-1$ and by (21) we have

$$\pi_s D_j^1 \underline{X} \cong \pi_s \underline{X}\ , \quad s \geq a \tag{22}$$

REMARK 16: A long exact sequence analogous to (21) can be obtained for singular homology theory (or any generalized homology theory) by defining

$$H_*(\underline{X}) = H_*(X, \bigcup_{j=1}^{a} X_j)\ .$$

Here $\underline{X} = (X;\{X_j\})$ is an (unpointed) (a+1)-ad, and to avoid technical difficulties we assume that $\underline{X}$ is either CW or open. The analogue of (21) for H_* is then obtained by an excision isomorphism from the long exact sequence of the triple

$$(X, \bigcup_{i=1}^{a} X_i, \bigcup_{i \neq j} X_i)\ .$$

Notice that it would not have been good to define $\pi_*(\underline{X}) = \pi_*(X, \bigcup_{j=1}^{a} X_j)$, because homotopy theory does not satisfy excision. However, we do have a map

$$\pi_s(\underline{X}) \longrightarrow \pi_s(X, \bigcup_{j=1}^{a} X_j) \ . \tag{23}$$

Namely, inclusion induces a map

$$\pi_s(\underline{X}) \to \pi_s\left(X; \left\{ \bigcup_{j=1}^{a} X_j \right\}_{i=1}^{a}\right)$$

and the right-hand side is isomorphic (by iterated use of Remark 15) to $\pi_s(X, \bigcup_{j=1}^{a} X_j)$. In the case $a=2$ the "Triad Connectivity Theorem" of Blakers and Massey ([A], p.105) gives a range of low dimensions in which excision <u>does</u> hold in homotopy theory, i.e., in which (23) is an isomorphism. We will cite a generalization to $(a+1)$-ads of the Blakers-Massey theorem, due to Barratt and Whitehead (Lemma 19 below) and use it to prove an even more general version (Lemma 18).

Recall that a map of spaces $X \to Y$ is a <u>weak homotopy equivalence</u> (w.h.e.) if it induces isomorphisms $\pi_s(X) \to \pi_s(Y)$ for all $s \geq 0$ and all basepoints. Let us call a map of $(a+1)$-ads $\left(X; \{X_j\}_{j=1}^{a}\right) \longrightarrow \left(Y; \{Y_j\}_{j=1}^{a}\right)$ a w.h.e. if for each set $S \subset \{1, \cdots a\}$ it maps X_S to Y_S by a w.h.e.

§I.B.3. Connectivity of (a+1)-ads.

Recall that a pair (X,Y) is called k-connected $(k\geq 0)$ if $\pi_0(Y) \longrightarrow \pi_0(X)$ is surjective and if for all basepoints in Y $\pi_s(X,Y)$ is trivial for all s such that $0<s\leq k$. We will generalize this notion to (a+1)-ads.

NOTATION. For the remainder of §I.B. $\underline{X}$ will be a CW or open (a+1)-ad $\left(X;\{X_j\}_{j=1}^{a}\right)$ with $a\geq 1$. "I" will denote the index set $\{1,\cdots a\}$. For $S\subset I$ we write

$$X_S = \bigcap_{j\in S} X_j$$

as before. We also write

$$X^S = \bigcup_{j\in S} X_j$$

and, for $\mathcal{S}\subset 2^I$,

$$^{\mathcal{S}}X = \bigcup_{S\in\mathcal{S}} X_S \ .$$

DEF 17. Let $a\geq 1$. Set $I = \{1,\cdots a\}$. Let $\underline{k} = \{k_T\}_{T\subset I}$ be a collection of integers ≥ 0 satisfying

$$k_T \leq \sum_j k_{T_j} \quad \text{whenever} \quad T \subset \coprod_j T_j \ . \tag{24}$$

A (CW or open) $(a+1)$-ad $\underline{X}$ is $\underline{k}$-connected if for each nonempty $T \subset I$ the pair

$$(X_{I-T}\ ,\ X_{I-T} \cap X^T)$$

is k_T-connected.

§I.B.4. "Homotopy Excision" for $(a+1)$-ads.

The main result of §I.B. is:

LEMMA 18. If a CW or open $(a+1)$-ad $\underline{X}$ $(a\geq 1)$ is $\underline{k}$-connected and $k_T \geq 2$ for all nonempty $T \subset I$, then $\pi_s(\underline{X}) = 0$ for $a\leq s\leq k_I$, for every basepoint in X_I.

The proof of Lemma 18 will be based on the following special case:

LEMMA 19. (Barratt-Whitehead [B-W] p.428). The conclusion of Lemma 18 holds if the additional hypothesis:

(25) $$X = \bigcup_{j \in I} X_{I-j}$$

is satisfied.

REMARK 20. Hypothesis (25) (In [B-W] it is called completeness) implies that $X_{I-T} \subset X^T$ for all $T \subset I$ with more than one element. Thus the definition of $\underline{k}$-connectedness becomes very redundant in this case.

REMARK 21. The real point of [B-W] is a computation of the first nonvanishing group $\pi_{1+\sum_{j=1}^{a} k_j}(\underline{X})$ in terms of the groups $\pi_{1+k_j}(X, X_j)$. However, we only need the vanishing statement.

REMARK 22. Barratt and Whitehead only deal with the case in which X is one-connected. However, the vanishing statement in general follows from the statement in that special case by replacing X by the universal cover of any component of X, and each X_j by its inverse image.

REMARK 23. Actually Barratt and Whitehead deal with CW $(a+1)$-ads. However, the CW version of Lemma 19 implies the open version; if $\underline{X}$ satisfies (25) and is open, then there exist a CW $(a+1)$-ad $\underline{X}'$ also satisfying (25) and a w.h.e. $\underline{X}' \to \underline{X}$. (Take $X' = |S.X|$, the geometric realization of the total singular complex. In X' set $X'_j = \bigcup_{i \neq j} |S.(\bigcap_{\ell \neq i} X_\ell)|$. Set $\underline{X}' = (X'; \{X'_j\}).)$.

In order to reduce Lemma 18 to Lemma 19 we need:

LEMMA 24. Let $\underline{X}$ and $\underline{k}$ be as in the hypothesis of Lemma 18. If $\underline{X}$ is $\underline{k}$-connected then for all $S \subset 2^I$ and nonempty $T \subset I$ the pair

$$({}^S X, {}^S X \cap X^T)$$

is k_T-connected.

PROOF of Lemma 24: First observe that "connectivity of pairs is preserved by pushout." That is, if $(A \cup B; A, B)$ is an open or CW triad and $(A, A \cap B)$ is k-connected (for some $k \geq 0$) then $(A \cup B, B)$ is k-connected. (By Remark 23 above it suffices to prove the CW case.

In that case it can be arranged by a (weak) homotopy equivalence of triads that only cells of dimension $>k$ are used in the pair $(A, A \cap B)$.)

<u>Step 1</u>. Without loss of generality X is 1-connected. (See Remark 22 above.)

<u>Step 2</u>. For any nonempty subsets S and S' of 2^I such that ${}^{S'}X \subset {}^{S}X$, the pair $({}^{S}X, {}^{S'}X)$ is 2-connected. To prove this, use induction on ${}^{S}X$ with respect to inclusion. Assume it is true for all smaller ${}^{S}X$.

We can certainly assume that ${}^{S'}X$ is strictly contained in ${}^{S}X$. In fact we can take ${}^{S'}X$ to be maximal in ${}^{S}X$, because if

$$ {}^{S'}X \subsetneq {}^{S''}X \subsetneq {}^{S}X $$

with $({}^{S}X, {}^{S''}X)$ 2-connected, then $({}^{S}X, {}^{S'}X)$ is also 2-connected since by induction on ${}^{S}X$ $({}^{S''}X, {}^{S'}X)$ is 2-connected.

<u>Case 1</u>. ${}^{S}X$ can be written as a union ${}^{S1}X \cup {}^{S2}X$ of strictly smaller sets. Then by maximality we have ${}^{S}X = {}^{Si}X \cup {}^{S'}X$ for either $i=1$ or $i=2$. Now by the "pushout" observation above it suffices to show that $({}^{Si}X, {}^{Si}X \cap {}^{S'}X)$ is 2-connected, which is true by induction on ${}^{S}X$.

<u>Case 2</u>. ${}^{S}X$ cannot be so written. Then ${}^{S}X = X_S$ for some $S \subset I$. Choose S maximal. Then we have

$$(26) \qquad X_S \supsetneq X_S \cap X^{I-S} .$$

For otherwise

$${}^{S}X = X_S = X_S \cap X^{I-S} = \bigcup_{T \supsetneq S} X_T ,$$

and since we are in Case 2 ${}^{S}X = X_T$ for some $T \supsetneq S$, contradicting maximality of S. On the other hand, we have

$$(27) \qquad {}^{S'}X \subset X_S \cap X^{I-S} ,$$

because for each $T \in S'$

$$X_T \subsetneq X_S \text{ , whence } T \not\subset S \text{ and}$$

$$X_T \subset X^{I-S} \text{ .}$$

Now (26) and (27) and the maximality of ${}^{S'}X$ in ${}^{S}X$ imply

$$(28) \qquad ({}^{S}X, {}^{S'}X) = (X_S, X_S \cap X^{I-S}) \text{ .}$$

This pair is k_{I-S}-connected, and so 2-connected, by hypothesis. ($I-S \neq \emptyset$, because otherwise ${}^{S}X = X_I$ has no proper subsets ${}^{S'}X$.)

Step 3. Each pair $({}^{S}X, {}^{S}X \cap X^T)$, $S \neq \emptyset$, $T \neq \emptyset$ is k_T-connected. Again use induction on ${}^{S}X$. Consider the same two cases.

Case 1. ${}^{S}X = {}^{S1}X \cup {}^{S2}X$ nontrivially. Set ${}^{S3}X = {}^{S1}X \cap {}^{S2}X$. Consider the 4-ad

$$\underline{Y} = ({}^{S}X; {}^{S1}X, {}^{S2}X, {}^{S}X \cap X^T) \text{ .}$$

The faces $D_3^0\underline{Y}$ and $D_3^1\underline{Y}$ have zero homology in all dimensions, by excision. Therefore $\underline{Y}$ has zero homology in all dimensions. The pairs $({}^{Si}X, {}^{Si}X \cap X^T)$ $(i=1,2,3)$ are k_T-connected by induction on ${}^{S}X$, and so have zero homology in dimensions $\le k_T$. It follows that the same is true of the homology of $({}^S X, {}^S X \cap X^T)$. But by Step 2 this pair is 2-connected, and by Steps 1 and 2 ${}^S X$ is 1-connected. It follows by the relative Hurewicz theorem that the pair is k_T-connected.

Case 2. As in Step 2 ${}^S X = X_S$, some S, and $X_S \supsetneq X_S \cap X^{I-S}$. If $S \cap T \neq \emptyset$ then there is nothing to prove. If $S \cap T = \emptyset$ then

$$X_S \cap X^T \subset X_S \cap X^{I-S} \subset X_S .$$

The small pair is k_T-connected by induction on ${}^S X$. The large pair is k_{I-S}-connected by hypothesis, hence k_T-connected by (24). Therefore the third pair is k_T-connected. QED.

PROOF of Lemma 18: We use induction with respect to a. If $a=1$ there is nothing to prove. Assume the Lemma for $(i+1)$-ads for all i such that $1 \le i < a$. Let $\underline{X}$ and $\underline{k}$ be as in the hypothesis of Lemma 18. Choose a basepoint in X_I.

List all of the subsets of $I = \{1, \cdots a\}$:

$$T_1, \cdots T_{2^a}$$

in such a way that

(29) $$1 \le i \le j \le 2^a \Rightarrow |T_i| \ge |T_j| .$$

Set

$$S^i = \{T_1, \cdots T_i\} , \quad \text{for} \quad 1 \le i \le 2^a .$$

Define

$$Y^i = {}^{S^i}X$$

$$\underline{Y}^i = (Y^i; Y^i_1, \cdots Y^i_a)$$

$$= (Y^i; Y^i \cap X_1, \cdots Y^i \cap X_a)$$

We will prove

$(30)_i$ $$\pi_s(\underline{Y}^i) = 0 \quad \text{for} \quad a \le s \le k_I$$

for $a+1 \leq i \leq 2^a$ by induction on i. Since $\underline{Y}^{2^a} = \underline{X}$ this will complete the proof of Lemma 18.

The proof in the case $i=a+1$ is a straightforward application of Lemma 19. Indeed, $Y^{a+1} = \bigcup_{j=1}^{a} X_{I-j}$, so that $\underline{Y}^{a+1}$ satisfies (25); and for each $j \in I$

$$(Y^{a+1}_{I-j}, Y^{a+1}_{I}) = (X_{I-j}, X_I)$$

is k_j-connected. Thus by Lemma 19 $(30)_{a+1}$ holds.

Now let $a+1 < i \leq 2^a$, and assume $(30)_{i-1}$. We will prove $(30)_i$ by defining an $(a+1)$-ad $\underline{Z}$ and a long exact sequence

$$\text{(31)} \qquad \cdots \to \pi_{a+1}(\underline{Y}^{i-1}) \to \pi_{a+1}(\underline{Y}^{i}) \to \pi_{a+1}(\underline{Z}) \to \pi_{a}(\underline{Y}^{i-1}) \to \pi_{a}(\underline{Y}^{i}) \to \pi_{a}(\underline{Z}) \quad .$$

Let $m = |T_i|$. Because $i>a+1$, we have

$$\text{(32)} \qquad 0 \leq m \leq a-2.$$

Without loss of generality (by permuting $\{1,\cdots a\}$) we can assume that $T_i = \{1,\cdots m\}$.

CLAIM 25. $Y^{i-1} = Y^i \cap X^{I-T_i}$.

PROOF of Claim: The inclusion

$$Y^{i-1} \subset Y^i$$

is obvious. To show

$$Y^{i-1} \subset X^{I-T_i}$$

assume $x \in Y^{i-1}$. Then $x \in X_{T_k}$ for some $k<i$. Now $T_k \neq T_i$, and by (29) $|T_k| \geq |T_i|$. Thus $T_k \not\subset T_i$. Choose $j \in T_k$, $j \notin T_i$. Then $x \in X_j \subset X^{I-T_i}$. On the other hand,

$$Y^i \cap X^{I-T_i} = (Y^{i-1} \cup X_{T_i}) \cap X^{I-T_i}$$

$$\subset Y^{i-1} \cup \bigcup_{j=m+1}^{a} X_{T_i \cup j} = Y^{i-1} . \quad \text{QED}$$

In particular we have

$$(33)_j \qquad Y^i_j \subset Y^{i-1}$$

for all $j>m$. Set

$$\underline{Z} = (Y^i;\ Y^i_1, \cdots Y^i_{a-1}, Y^{i-1}).$$

Then $(33)_a$ establishes the long exact sequence (31).

It remains to show that $\pi_s(\underline{Z}) = 0$ for $a \le s \le k_I$. Define an $(m+2)$-ad $\underline{W}$ by

$$\underline{W} = (W; W_1, \cdots W_{m+1})$$

$$= (Y^i; Y^i_1, \cdots Y^i_m, Y^{i-1}).$$

The relations $(33)_j$, $m<j<a$, imply that $\pi_s(\underline{W}) \cong \pi_s(\underline{Z})$ for all $s \ge a$. We will show that $\pi_s(\underline{W}) = 0$ for $m+1 \le s \le k_I$, using the inductive hypothesis with respect to a. By (32) $m+1 < a$, so that the inductive hypothesis applies. Set $\underline{k}' = \{k'_{T'}\}$, where we define integers $k'_{T'}$ indexed by the subsets $T' \subset \{1, \cdots m+1\}$ by

$$k'_{T'} = k_{T'} \qquad \text{for} \qquad T' \subset \{1, \cdots m\}$$

$$k'_{T \cup \{m+1\}} = k_{T \cup \{m+1, \cdots a\}} \text{ for } T \subset \{1, \cdots m\}.$$

Then $\underline{W}$ is $\underline{k}'$-connected. Indeed, let T' be a nonempty subset of $I' = \{1, \cdots m+1\}$. There exists a nonempty subset S of 2^I such that

$$W_{I'-T'} = {}^{S}X .$$

If $m+1 \notin T'$ then

$$(W_{I'-T'}, W_{I'-T'} \cap W^{T'}) = ({}^{S}X, {}^{S}X \cap X^{T'})$$

is $k_{T'}$-connected by Lemma 24, hence $k'_{T'}$-connected. If $m+1 \in T'$, then write $T' = T \amalg \{m+1\}$. We have

$$(W_{I'-T'}, W_{I'-T'} \cap W^{T'}) = ({}^{S}X, {}^{S}X \cap X^{T\cup\{m+1,\cdots a\}})$$

is $k_{T\cup\{m+1,\cdots a\}}$-connected, hence $k'_{T'}$-connected. (The equality above uses that $W_{m+1} = W \cap X^{\{m+1,\cdots a\}}$, which is true by Claim 25.) It follows that $\pi_s \underline{W} = 0$ for $m+1 \le s \le k'_{I'} = k_I$, which completes the proof of Lemma 18. QED

§I.C. Sunny Collapsing.

Eventually it will be necessary, in proving Theorem D, to construct certain fibered isotopies of concordances of P in N. This section provides a way of producing such isotopies. Given a fibered concordance F of P in N and given a collection of data called a "smooth fibered sunny collapse"* we construct a fibered isotopy F^u with $F^0 = F$. This technique will be used in Chapter III.

Suppose that F is a smooth fibered concordance over D^s

$$I \times P \times D^s \xrightarrow{F = (h, f, p_3)} I \times N \times D^s$$

(See §I.A.1 for definitions.)

Suppose that ϕ^u and ψ^u are smooth homotopies

$$\begin{aligned} \phi^u &: P \times D^s \longrightarrow (0,1] \\ \psi^u &: N \times D^s \longrightarrow (0,1] \end{aligned} \qquad 0 \le u \le 1$$

which satisfy:

(34) $\phi^u(x,y) = 1$ if $u=0$ or $x \in \partial P$

(35) $\psi^u(z,y) = 1$ if $u=0$ or $z \in \partial N$

(36) $h(t,x,y) = \psi^u(f(t,x,y),y)$ if $t=\phi^u(x,y)$

(37) $h(t,x,y) < \psi^u(f(t,x,y),y)$ if $t<\phi^u(x,y)$

---

*The name is intended to suggest an analogy with the PL technique called "sunny collapsing." See [Mil].

(38) $\frac{\partial}{\partial t}(h(t,x,y) - \psi^u(f(t,x,y),y))>0$ for any (u,t,x,y) such that $t = \phi^u(x,y)$.

Then the smooth homotopy

$$I\times P\times D^s \xrightarrow{F^u} I\times N\times D^s \qquad 0 \le u \le 1$$

given by the formula:

(39)
$$F^u(t,x,y) = \left(\frac{h(t\phi^u(x,y),x,y)}{\psi^u(f(t\phi^u(x,y),x,y),y)}, f(t\phi^u(x,y),x,y),y\right)$$

is easily seen to be a smooth fibered isotopy of concordances with $F^0 = F$.

Here is a restatement of (36), (37), and (38) in geometric terms: Let $G(\phi^u)$ and $G(\psi^u)$ be the graphs

$$G(\phi^u) = \left\{(t,x,y) \in I\times P\times D^s \;\middle|\; t = \phi^u(x,y)\right\}$$

$$G(\psi^u) = \left\{(t,x,y) \in I\times N\times D^s \;\middle|\; t = \psi^u(x,y)\right\}$$

and let $M(\phi^u)$ and $M(\psi^u)$ be the regions bounded above by the graphs:

$$M(\phi^u) = \left\{ (t,x,y) \in I\times P\times D^s \;\middle|\; t < \phi^u(x,y) \right\}$$

$$M(\psi^u) = \left\{ (t,x,y) \in I\times N\times D^s \;\middle|\; t < \psi^u(x,y) \right\}$$

Then (36) says

$$F(G(\phi^u) \subset G(\psi^u) \quad \text{for all} \quad u,$$

and (37) says

$$F(M(\phi^u) \subset M(\psi^u) \quad \text{for all} \quad u\ .$$

In view of (36) and (37), (38) says

> F is transverse to $G(\psi^u)$ at every point in $G(\phi^u)$, for all u .

Visually, think of u as time. Think of $G(\phi^u)$ as a moving hypersurface in $I\times P\times D^s$ which begins at the "top" $(G(\phi^\circ) = 1\times P\times D^s)$ and moves down*:

---

*(34)-(38) do not actually require that ϕ^u and ψ^u be non-increasing as functions of u, but this added assumption may make visualizing a little easier, and in practice will usually hold.

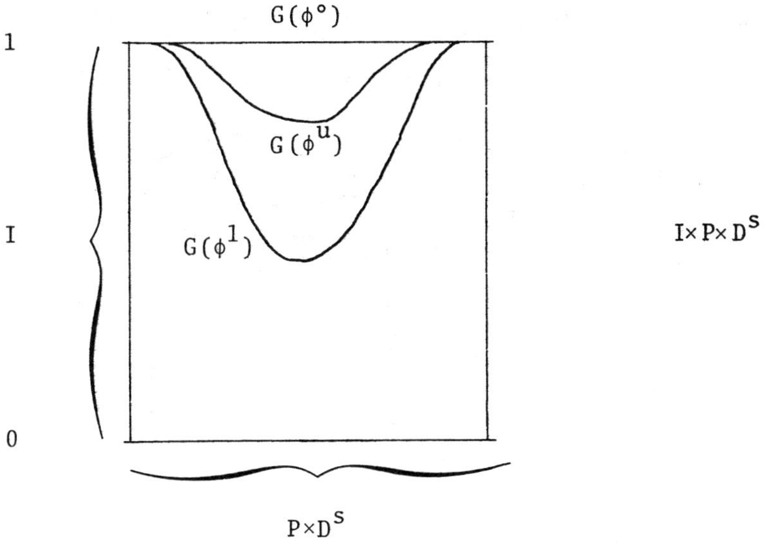

while at the same time in $I \times N \times D^s$ another family $G(\psi^u)$ moves down from $1 \times N \times D^s$:

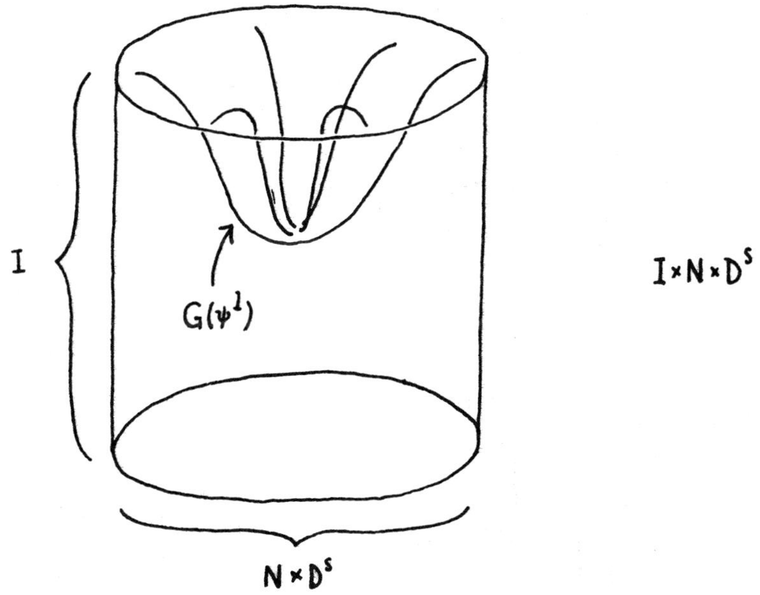

in such a way that the image $F(I\times P\times D^s)$

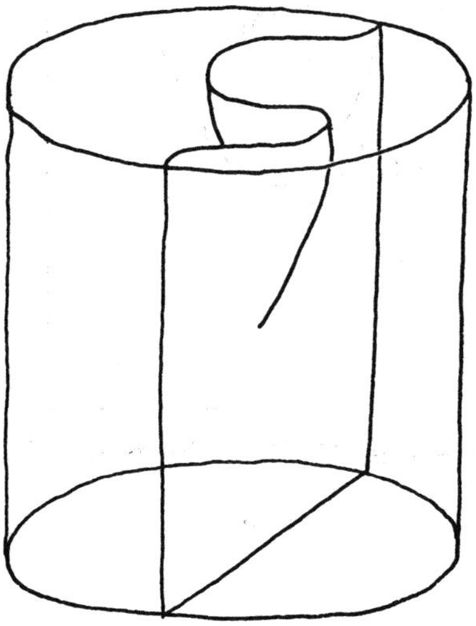

is cut transversely by each $G(\psi^u)$, precisely in $F(G(\phi^u))$.

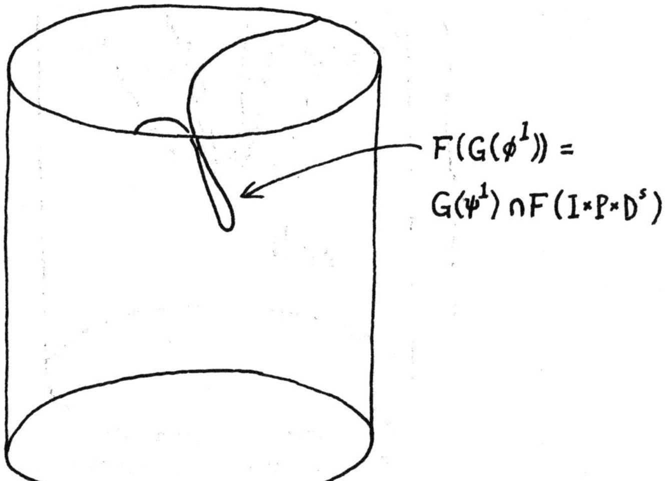

The map F^u defined by (39) is just made from F by making the diagram commute:

$$\begin{array}{ccc} I\times P\times D^s & \xrightarrow{F^u} & I\times N\times D^s \\ \| \wr & & \| \wr \\ \overline{M(\phi^u)} & & \overline{M(\psi^u)} \\ \downarrow & & \downarrow \\ I\times P\times D^s & \xrightarrow{F} & I\times N\times D^s \end{array}$$

where the isomorphisms are determined by ϕ^u and ψ^u. (See also the pictures in §D.5. of the Introduction.)

DEFINITION 26. Let X be a manifold. A point $x \in I\times X$ is <u>below</u> $y \in I\times X$ if $p_1(x) < p_1(y)$ and $p_2(x) = p_2(y)$. A vector $v \in T(I\times X)$ is <u>vertical</u> if $(Dp_2)\cdot v = 0$, and is <u>upward vertical</u> if in addition $(dp_1)\cdot v > 0$.

Exercise: If ϕ^u and ψ^u satisfy (34) through (38) then ϕ^u must satisfy (40) and (41):

(40) $0 \le t \le \phi^u(x,y) \implies F(\phi^u(x',y),x',y)$ is not below $F(t,x,y)$, for any $(u,t,x,x',y) \in I\times I\times P\times P\times D^s$

(41) If v is a tangent vector to $I\times P\times D^s$ at a point (t,x,y) such that $t = \phi^u(x,y)$, and if $d(\phi^u \circ p_{2,3})\cdot v \geq dp_1\cdot v$, then $DF\cdot v$ is not upward vertical (in $I\times N\times D^s$) .

What (40) says, more geometrically, is that no point in $F(G(\phi^u))$ is below a point in $\overline{F(M(\phi^u))}$. That is, if x and y in $I\times P\times D^s$ are such that $F(x)$ is below $F(y)$, then the moving surface $G(\phi^u)$ may not sweep through x unless it has already swept through y.

Similarly (41) says that no inward-pointing vector of $\overline{F(M(\phi^u))}$ at a point in $F(G^u)$ may be upward vertical. That is, if x in $I\times P\times D^s$ is such that the tangent space of $F(I\times P\times D^s)$ at $F(x)$ contains the vertical direction in the tangent space of $I\times N\times D^s$, then the surface $G(\phi^u)$ is not allowed to sweep through x in such a way that the image $F(G(\phi^u))$ moves upward through $F(x)$.

We restate (40) and (41) yet again: Think of the sun shining down into $I\times N\times D^s$ from ∞ in $\mathbb{R}_+\times N\times D^s$. Inside $I\times N\times D^s$ is the submanifold $F(I\times P\times D^s)$. As u runs from 0 to 1, $\overline{F(M(\phi^u))}$ melts away, starting as $\overline{F(M(\phi^0))} = \overline{F(I\times P\times D^s)}$ and ending as $\overline{F(M(\phi^1))}$. Statement (40) asserts that the surface $F(G(\phi^u))$ where the melting is taking place at time u is always in the sunshine, i.e. is not in the shadow of $\overline{F(M(\phi^u))}$. Statement (41) adds that this rule is to be interpreted in the following extended sense: a point in $F(G(\phi^u))$ is not considered to be truly "in the sunshine" if it is struck by a ray of light which points in an outward tangent direction of $F(M(\phi^u))$, thus:

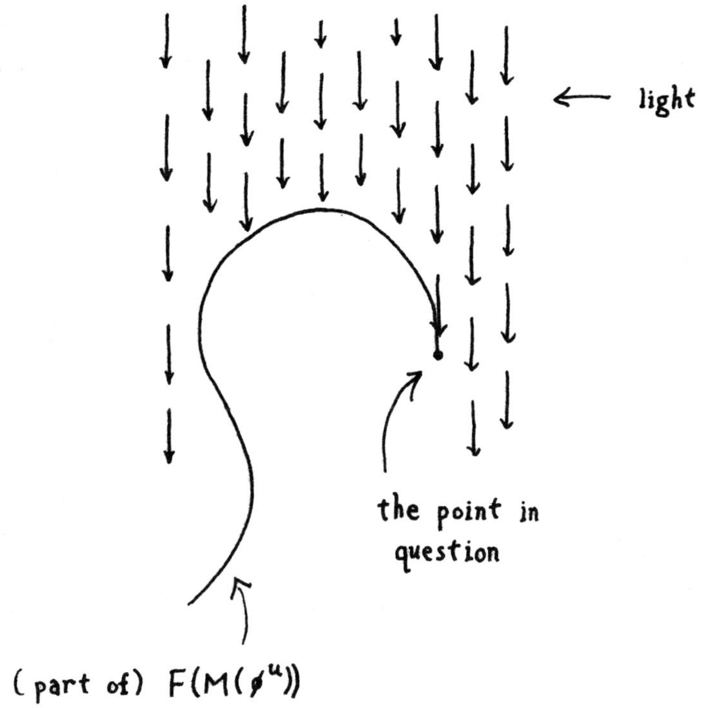

The implication (34)-(38) $\Longrightarrow$ (40)-(41) will not be needed here. The following converse will be needed. It is a statement which becomes very plausible — almost obvious — if one thinks about it awhile. However, its proof is a little long.

LEMMA 27. If F is a fibered concordance and ϕ^u is a smooth homotopy

$$\phi^u\colon P\times D^s \longrightarrow (0,1] \qquad 0 \le u \le 1$$

satisfying (34), (40), and (41), then there is a smooth homotopy

$$\psi^u\colon N\times D^s \longrightarrow (0,1] \qquad 0 \le u \le 1$$

which together with ϕ^u satisfies (35), (36), (37), and (38).

PROOF of Lemma 27. The problem can be reduced to a local one by using a partition of unity. Indeed, suppose $\{U_\alpha\}$ is an open cover of $I\times N\times D^s$ and for each α

$$\psi_\alpha\colon U_\alpha \longrightarrow (0,1]$$

is a smooth function. Write

$$\psi_\alpha^u(z,y) = \psi_\alpha(u,z,y) \quad \text{for} \quad (u,z,y) \in U_\alpha\ .$$

Suppose that ψ_α satisfies (35)-(38) whenever they make sense. Then choose a partition of unity

$$\{\mu_\alpha : I\times N\times D^s \longrightarrow [0,1]\}$$

subordinate to $\{U_\alpha\}$, and set

$$\psi^u(z,y) = \sum_{\substack{\alpha \\ (u,z,y)\in U_\alpha}} \mu_\alpha(u,z,y)\psi^u_\alpha(z,y) \quad \text{for} \quad (u,z,y) \in I\times N\times D^s$$

We check (35) through (38) for ψ^u. Both (35) and (36) follow immediately from the corresponding statements for the ψ^u_α. For (37), suppose $t < \phi^u(x,y)$. Then

$$\begin{aligned}\psi^u(f(t,x,y),y) &= \sum_\alpha \mu_\alpha(u,f(t,x,y),y)\ \psi^u(f(t,x,y),y) \\ &> \sum_\alpha \mu_\alpha(u,f(t,x,y),y)h(t,x,y)\end{aligned}$$

(each term in the first sum is greater than or equal to the corresponding term in the second, because each ψ^u_α satisfies (37), and for at least one term the inequality is strict since $\mu_\alpha > 0$)

$$= h(t,x,y) .$$

For (38), write

$$\frac{\partial}{\partial t}(h(t,x,y)-\Psi^u(f(t,x,y),y))$$

$$= \frac{\partial}{\partial t}\sum_\alpha \mu_\alpha(u,f(t,x,y),y)(h(t,x,y)-\psi^u_\alpha(f(t,x,y),y))$$

$$= \sum_\alpha \mu_\alpha(u,f(t,x,y),y)\frac{\partial}{\partial t}(h(t,x,y)-\psi^u_\alpha(f(t,x,y),y))$$

(for any (u,t,x,y) such that $t = \phi^u(x,y)$, because ψ^u_α satisfies (36). The last expression is positive because ψ^u_α satisfies (38). This proves (38).

Now to solve the local problem. Let

$$(u_0,z_0,y_0) \in I\times N\times D^s .$$

The problem is to find a neighborhood U_α of (u_0,z_0,y_0) and a smooth function

$$\psi_\alpha\colon U_\alpha \longrightarrow (0,1]$$

such that ψ_α satisfies (35) through (38) whenever

$$(42)\quad (u,f(t,x,y),y) \in U_\alpha \quad (\text{or, for } (35),\quad (u,z,y) \in U_\alpha\ ) .$$

In fact it will be enough if ψ_α satisfies (35) and (36) whenever (42) holds, and satisfies (37) and (38) whenever

$$(43)\qquad (u,f(t,x,y),y) = (u_0,z_0,y_0) .$$

For then in the set

$$\{(u,t,x,y) \mid (42)\}$$

the following function:

$$\begin{cases} \dfrac{h(t,x,y) - \psi_\alpha(u,f(t,x,y),y)}{t - \phi^u(x,y)} & \text{if} \quad t \neq \phi^u(x,y) \\ \dfrac{\partial}{\partial t}\,(h(t,x,y) - \psi_\alpha(u,f(t,x,y),y) & \text{if} \quad t = \phi^u(x,y) \end{cases}$$

is smooth (by (36) applied to ψ_α) and in particular continuous. By assumption the function is negative in the set

$$\{(t,x,y) \mid (43) \quad \text{and} \quad t \leq \phi^u(x,y)\} .$$

Therefore it is negative in some neighborhood of this set. It follows that for some neighborhood U'_α of (u_0,z_0,y_0) in U_α, the function is negative wherever

$$(u,f(t,x,y),y) \in U'_\alpha \quad \text{and} \quad t \le \phi^u(x,y) .$$

That is, $\psi_\alpha\big|_{U'_\alpha}$ satisfies (37) and (38).

Now to find ψ_α and U_α.

Case 1. There is no $x \in P$ such that

$$f(\phi^{u_0}(x,y_0),x,y_0) = z_0 .$$

Choose U_α small enough so that

$$x \in P,\ (u,z,y) \in U_\alpha \implies f(\phi^u(x,y),x,y) \ne z .$$

Set

$$\psi^u_\alpha(z,y) = 1 \quad \text{for all} \quad (u,z,y) \in U_\alpha .$$

Then for ψ^u_α (35) is trivial, (36) is vacuous, (37) is easy ($t < \phi^u(x,y) \Rightarrow t < 1 \implies h(t,x,y) < 1$), and (38) is vacuous.

Case 2. For some $x_0 \in P$

$$f(\phi^{u_0}(x_0,y_0),x_0,y_0) = z_0 .$$

Consider the smooth map

$$E: \ I\times P\times D^s \longrightarrow I\times N\times D^s$$

$$E(u,x,y) = (u,f\ (\phi^u(x,y),x,y),y)$$

(40) and (41) imply that E is an injection and an immersion, respectively. Indeed, first assume

$$E(u',x',y') = E(u,x,y)$$

$$(u',x',y') \neq (u,x,y) .$$

That is,

$$u' = u,\ y' = y,\ x' \neq x$$

$$f(\phi^u(x',y),x',y) = f(\phi^u(x,y),x,y) .$$

Now

$$h(\phi^u(x',y),x',y) \neq h(\phi^u(x,y),x,y)$$

because F is injective. Without loss of generality

$$h(\phi^u(x',y),x',y) < h(\phi^u(x,y),x,y)$$

That is,

$$F(\phi^u(x',y),x',y) \text{ is below } F(\phi^u(x,y),x,y) .$$

Then (40) implies

$$\phi^u(x,y) > \phi^u(x,y) ,$$

a contradiction. Therefore E is an injection. Next assume

$$(44) \quad \begin{cases} DE\cdot w = 0 , \\ w \neq 0 \quad \text{a tangent vector to } I\times P\times D^s \text{ at } (u_1,x_1,y_1) . \end{cases}$$

(44) implies $(Dp_1)\cdot w = 0$ and $(Dp_3)\cdot w = 0$, so that $(Dp_2)\cdot w \neq 0$ (since $w \neq 0$). Set

$$v = D((u,x,y) \longmapsto (\phi^u(x,y),x,y))\cdot w .$$

Then

$$d(\phi^{u_1}\circ p_{2,3})\cdot v = d(\phi^{u_1}\circ p_{2,3})\cdot w$$

$$= (dp_1)\cdot v$$

because $(dp_1)\cdot w = 0$. Also

$$(Dp_2)\cdot v = (Dp_2)\cdot w \neq 0, \text{ hence } v \neq 0 .$$

But

$$DF\cdot v = D((u,x,y) \longmapsto (h(\phi^u(x,y),x,y),f(\phi^u(x,y),x,y),y))\cdot w$$

is vertical (because $DE\cdot w = 0$) and nonzero (because F is an embedding and $v \neq 0$). Therefore either $\pm v$ violates (41). This contradiction proves that E is an immersion. In fact, E is clearly a proper fibered embedding of P in N over $I\times D^s$.

By assumption (in Case 2) (u_0,z_0,y_0) is in the image of E. Because E is an embedding, there is certainly a function ψ_α defined near (u_0,z_0,y_0) such that

$$\psi_\alpha(E(u,x,y) = h(\phi^u(x,y),x,y) ,$$

that is, such that ψ_α satisfies (36). Also, because E is a proper embedding and

$$h(\phi^u(x,y),x,y) = 1 \quad \text{when} \quad u = 0 \quad \text{or} \quad x \in \partial P ,$$

it is possible to arrange for (35) to hold for ψ_α in some neighborhood of (u_0,z_0,y_0) at the same time. (37) will hold automatically for (u,t,x,y) if (43) holds. For suppose (43) and

$$t < \phi^u(x,y) . \tag{45}$$

Thus $u = u_0$, $y = y_0$, and

$$f(t,x,y_0) = z_0 = f(t_0,x_0,y_0) . \tag{46}$$

Now

$$(t,x,y_0) \neq (\phi^{u_0}(x_0,y_0),x_0,y_0)$$

because if $x = x_0$ then

$$t < \phi^u(x,y) = \phi^{u_0}(x_0,y_0) .$$

Thus in view of (46) we have

$$h(t,x,y_0) \neq h(\phi^{u_0}(x_0,y_0),x_0,y_0);$$

and (40), with (u_0,t,x,x_0,y_0) in place of (u,t,x,x',y), says (again in view of (46)), that

$$\begin{aligned} h(t,x,y_0) &< h(\phi^{u_0}(x_0,y_0),x_0,y_0) \\ &= \psi_\alpha^{u_0}(f(\phi^{u_0}(x_0,y_0)),y_0) \end{aligned}$$

as (37) requires.

Finally, ψ_α should be such that (38) holds whenever (43) holds. Suppose (43) and $t = \phi^u(x,y)$. Then

$$(u,x,y) = (u_0,x_0,y_0) .$$

Define a tangent vector w to $I\times N\times D^s$ at (u_0,x_0,y_0) by

$$w = D((t,x,y) \longmapsto (u_0,f(t,x,y),y)).\frac{\partial}{\partial t}$$

where $\frac{\partial}{\partial t}$ is the standard upward vertical vector field in $I\times P\times D^s$ ("partial derivative with respect to the first coordinate"), evaluated in this case at (t_0,x_0,y_0). What is needed for (38) is that

$$(47) \qquad d\psi_\alpha \cdot w < \frac{\partial h}{\partial t} \quad \text{at} \quad (u_0, z_0, y_0) \; .$$

If w is not tangent to the submanifold $E(IxPxD^s)$, then this is easy to achieve while also making sure of (35), (36) and (37). (In this case we are merely specifying one normal derivative of ψ_α at (u_0, z_0, y_0) in addition to its values on the submanifold.) If w is tangent to $E(IxPxD^s)$, say $w = DE \cdot v'$, then apply (41) to the vector

$$v \underset{def}{=} D((u,x,y) \longmapsto (\phi^u(x,y), x, y)) \cdot v' - \frac{\partial}{\partial t} \; .$$

One sees that $(DF) \cdot v$ is not upward vertical. Since it is vertical and is not zero, $(dh) \cdot v$ must be negative. That is,

$$d((u,x,y) \longmapsto h(\phi^u(x,y), x, y)) \cdot v' < \frac{\partial h}{\partial t}(t_0, z_0, y_0) \; .$$

That is, (47) is true automatically in this case. This completes the proof of Lemma 27. QED

DEFINITION 28. Let

$$I \times P \times D^s \xrightarrow{F = (h,f,p_3)} I \times N \times D^s$$

be a fibered concordance of P in N over D^s. A smooth <u>fibered sunny collapse</u> (or briefly a sunny collapse) relative to F is a smooth homotopy of functions

$$\phi^u \colon P \times D^s \longrightarrow (0,1] \ , \qquad 0 \le u \le 1$$

satisfying (34), (40), and (41).

To summarize the results of §I.C. so far:

LEMMA 29. Let

$$I \times P \times D^s \xrightarrow{F = (h,f,p_3)} I \times N \times D^s$$

be a fibered concordance. Let ϕ^u be a sunny collapse relative to F. Then there exists a smooth fibered isotopy of concordances $F^u = (h^u, f^u, p_3)$ with $F^0 = F$, such that for all $(u,t,x,y) \in I \times I \times P \times D^s$

$$f^u(t,x,y) = f^0(t\phi^u(x,y),x,y) \quad . \tag{48}$$

In practice when a sunny collapse ϕ^u is constructed it will often have the form

$$\phi^u(x,y) = 1 - u(1-\phi(x,y))$$

where $\phi(=\phi^1)$ is some function

$$\phi\colon P\times D^s \longrightarrow (0,1] .$$

For convenience, here are conditions on ϕ which suffice to make ϕ^u a sunny collapse.

LEMMA 30. Let F be a fibered concordance of P in N. Let

$$\phi\colon P\times D^s \longrightarrow (0,1]$$

be a smooth function. Assume

(49) $\phi(x,y) = 1$ if $x\in\partial P$

(50) If $F(t,x,y)$ is below $F(t',x',y)$, then either $\phi(x,y) > t$ or $(1-t)(1-\phi(x',y)) > (1-t')(1-\phi(x,y))$

(51) If $\phi(x_0,y_0) \leq t_0 < 1$ and $DF\cdot v = \frac{\partial}{\partial t}$, where v is some tangent vector of $I\times P\times D^s$ at (t_0,x_0,y_0), then $(d\phi)\cdot(Dp_{2,3})\cdot v < (\frac{1-\phi(x_0,y_0)}{1-t_0})(dp_1)\cdot v$.

Then the homotopy ϕ^u defined by

$$\phi^u(x,y) = 1-u(1-\phi(x,y))$$

is a sunny collapse with respect to F.

PROOF of Lemma 30: (34), (40), and (41) must be checked. (34) obviously follows from (49). (40) will follow from (50). Suppose (40) fails. Thus there exist

$$(t,x,y) \in I\times P\times D^s$$
$$(t',x',y) \in I\times P\times D^s$$
$$u \in [0,1]$$

such that

(52) $F(t,x,y)$ is below $F(t',x',y)$

(53) $1-u(1-\phi(x,y)) = t$

(54) $$1-u(1-\phi(x',y)) \geq t'$$

By (50) either

(55) $$\phi(x,y) > t$$

or

(56) $$(1-t)(1-\phi(x',y)) > (1-t')(1-\phi(x,y))$$

But (55) implies

$$1-u(1-\phi(x,y)) \geq \phi(x,y) > t ,$$

contradicting (53). On the other hand (52) implies $t < 1$, which by (53) implies $1-\phi(x,y) > 0$. Solve (53) for u, substitute the resulting value of u in (54), and multiply both sides by $1-\phi(x,y)$. This contradicts (56), proving (40).

Finally, (41) will follow from (51). Suppose (41) fails. Then there exist

$(t_0,x_0,y_0) \in I\times P\times D^s$

v tangent to $I\times P\times D^s$ at (t_0,x_0,y_0) , and

$u \in [0,1]$

such that

$$t_0 = \phi^u(x_0,y_0) \tag{57}$$

$$DF\cdot v = \frac{\partial}{\partial t} \tag{58}$$

$$d((t,x,y) \longmapsto \phi^u(x,y)-t)\cdot v \geq 0 \tag{59}$$

Suppose first that $t_0 = 1$. (58) implies that $(dh)\cdot v = 1$, which (since dh is a positive multiple of dp_1 at every point on $1\times P\times D^s$) implies

$$(dp_1)\cdot v > 0 . \tag{60}$$

But ϕ^u has a maximum at (x_0,y_0), since by (57) it takes the value 1 there. Therefore (59) becomes

$$-(dp_1)\cdot v \geq 0 ,$$

contradicting (60).

Therefore $t_0 < 1$. Now (51) applies. But (59) says

$$u(d\phi)(Dp_{2,3})\cdot v - (dp_1)\cdot v \geq 0 ,$$

that is,

$$(d\phi)(Dp_{2,3})\cdot v \geq \left(\frac{1-\phi(x_0,y_0)}{1-t_0} \right)(dp_1)\cdot v \quad ,$$

contradicting (51). This proves (41) and so completes the proof of Lemma 30. QED.

§I.D. Multijets and General Position.

Multijets provide a good context in which to state and prove many "general position" theorems in differential topology. The purpose of this section is to quote a very general theorem of that kind (Lemma 32) and deduce a variant of it which is especially adapted to the study of fibered concordances (Lemma 34).

For a thorough treatment of multijets see [Ma2] p.302 and [Ma1] pp.257-258; [Hi] Chapter 2 may also be helpful. Recall roughly what the set-up is: Suppose X and Y are two smooth manifolds and $m\geq 0$ is an integer. Two smooth maps

$$X \underset{g}{\overset{f}{\rightrightarrows}} Y$$

are said to agree to order m at $x \in X$ if $f(x) = g(x)$ and if (for one, hence any, choice of coordinate charts for X near x and Y near $f(x)$) f and g have the same Taylor polynomial of degree m at x. An m-jet from X to Y consists of a point $x \in X$ (the source of the jet) together with an equivalence class of maps f from X to Y under the relation of agreeing to order m at x. The point $f(x)$ (for any f in the class) is called the target of the jet. The set of all m-jets from X to Y is a manifold in a natural way. It is denoted $J^m(X,Y)$. The map

$$J^m(X,Y) \xrightarrow{(s,t)} X \times Y$$

taking a jet to its source and target is a smooth bundle projection. Every smooth map f from X to Y induces a smooth section

$$X \xrightarrow{j^m(f)} J^m(X,Y)$$

of s, namely

$$j^m(f)(x) = (x, \text{equivalence class of } f \text{ with respect to } x).$$

Multijets are a simple generalization of jets. If X, Y, and m are as above and $r \geq 1$ is an integer then an (r,m)-multijet from X to Y is an element of the manifold

$$ {}_rJ^m(X,Y) \underset{\text{def}}{=} \left\{ (z_1,\cdots z_r) \in (J^m(X,Y))^r \;\middle|\; 1 \leq i < j \leq r \Longrightarrow sz_i \neq sz_j \right\} . $$

Define source and target maps

$$ {}_rJ^m(X,Y) \xrightarrow{(s,t)} X^{(r)} \times Y^r $$

by

$$ s(z_1,\cdots z_r) = (sz_1,\cdots sz_r) $$

$$ t(z_1,\cdots z_r) = (tz_1,\cdots tz_r) . $$

Again each $f\colon X \longrightarrow Y$ induces a section

$$ X^{(r)} \xrightarrow{{}_rj^m(f)} {}_rJ^m(X,Y) $$

of s, namely

$$ {}_rj^m(f)(x_1,\cdots x_r) = (j^m(f)(x_1),\cdots j^m(f)(x_r)) . $$

There are obvious smooth maps (bundle projections, in fact)

$$p_m^{m'} : {}_rJ^{m'}(X,Y) \longrightarrow {}_rJ^m(X,Y), \qquad 0 \le m \le m' < \infty$$

It is sometimes convenient to speak of jets and multijets of infinite order. Thus one sets

$${}_rJ^\infty(X,Y) = \lim_{\underset{m}{\longleftarrow}} {}_rJ^m(X,Y) \quad ,$$

the inverse limit being taken with respect to the maps $p_m^{m'}$. (Of course, ${}_rJ^\infty(X,Y)$ is not a manifold. We will never even have to consider it as more than a set.) There are obvious source and target maps

$${}_rJ^\infty(X,Y) \xrightarrow{(s,t)} X^{(r)} \times Y^r \quad .$$

Let $C^\infty(X,Y)$ be the space of all smooth maps from X to Y. The topology here is the "strong Whitney topology". If C is a closed subset of X and f_0 is an element of $C^\infty(X,Y)$, define

$$C^\infty(X,Y;f_0,C) = \left\{ f \in C^\infty(X,Y) \;\middle|\; f|_C = f_0|_C \right\}$$

with the relative topology.

LEMMA 31. For any X, Y, f_0, and C as above, $C^\infty(X,Y;f_0,C)$ is a Baire space. (That is, in this space every countable intersection of dense open subsets is dense.)

PROOF of Lemma 31: Mather ([Ma2] Prop. 3.1 p.309) proves it in the case in which C, ∂X, and ∂Y are empty. The method used there can be easily adapted to prove the general statement. QED.

LEMMA 32. (Relative Multijet Transversality Theorem). Let X, Y, f_0, and C be as above. Let W be a submanifold of ${}_rJ^m(X,Y)$ for some $r \geq 1$ and (finite) $m \geq 0$. Then the set

$$\left\{ f \in C^\infty(X,Y;f_0,C) \;\middle|\; {}_rj^m(f|_{X-C}) \pitchfork W \right\}$$

is a countable intersection of dense open subsets of $C^\infty(X,Y;f_0,C)$.

PROOF: Again see [Ma2] for a generalizable proof of a special case (Prop. 3.3, p.310). QED

It is not difficult to deduce a variant of Lemma 32 for parametrized families of maps

$$f_y\colon X \longrightarrow Y \quad , \quad y \in M$$

with a smooth manifold M as parameter space. (It is understood that $f_y(x)$ depends smoothly on $(x,y) \in X\times M$.) Write

$$f_y(x) = f(x,y) \quad .$$

Given such a family f, one can form an associated multijet map

$${}_r\tilde{j}^m(f)\colon X^{(r)} \times M \longrightarrow {}_rJ^m(X,Y)$$

for any $r \geq 1$ and $m \geq 0$ by setting

$${}_r\tilde{j}^m(f)(x_1,\cdots x_r,y) = {}_rj^m(f_y)(x_1,\cdots x_r)$$

The parametrized version of Lemma 32 reads:

LEMMA 33. Let X, Y, r, m, W, and C be as in Lemma 32. Let M be a manifold. Then for any $f_0 \in C^\infty(X\times M, Y)$ the set

$$\left\{ f \in C(X\times M, Y; f_0, C\times M) \;\middle|\; {}_r\tilde{j}^m\left(f\Big|_{(X-C)\times M}\right) \pitchfork W \right\}$$

is a countable intersection of dense open subsets of $C^\infty(X\times M, Y; f_0, C\times M)$.

PROOF of Lemma 33: The plan is to make Lemma 33 into a special case of Lemma 32 by expressing ${}_r\tilde{j}^m(f)$ in terms of ${}_rj^m(f)$. Define

$${}_rJ^m(X\times M, Y) \xrightarrow{\ \Psi\ } J^m(X,Y)^r \times M^r$$

by

$$\Psi(z_1, \cdots z_r) = ((z_1 \cdot j^m(i_{p_2(sz_1)})(p_1(sz_1)), \cdots$$

$$z_r \cdot j^m(i_{p_2}(sz_r)(p_1(sz_r))),$$

$$(p_2(sz_1), \cdots p_2(sz_r))) \ ,$$

where p_1 and p_2 are the projections

$$X\times M \longrightarrow X$$
$$X\times M \longrightarrow M$$

and for each $y \in M$ i_y is the injection

$$X \longrightarrow X\times M \quad .$$
$$i_y(x) = (x,y)$$

Consider the diagram

$$\begin{array}{c}
X^{(r)} \times M \\
\big\downarrow e \\
(X\times M)^{(r)} \\
\big\downarrow {}_rj^m(f) \\
{}_rJ^m(X\times M, Y) \\
\big\downarrow \Psi \\
J^m(X,Y)^r \times M^r \\
\swarrow p_1 \qquad\qquad p_2 \searrow \\
J^m(X,Y)^r \qquad\qquad\qquad M^r
\end{array}$$

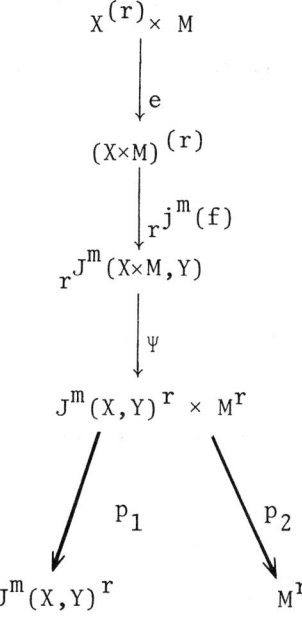

where e is the obvious embedding

$$e(x_1,\cdots x_r,y) = ((x_1,y),\cdots(x_r,y)) \quad .$$

Notice that

(61) $$p_1\cdot\Psi\cdot{}_rj^m(f) = {}_r\tilde{j}^m(f) \quad (\text{for any } f)$$

(62) Ψ is a bundle projection over its image, which is an open subset of its range

(63) $p_2\cdot\Psi\cdot{}_rj^m(f)$ is transverse to the diagonal $\Delta \subset M^r$, and the inverse image of Δ is the image of e .

Now set

$$W' = \Psi^{-1}(W\times\Delta).$$

W' is a submanifold by (62). Moreover

$$\begin{aligned}{}_rj^m(f) \pitchfork W' &\iff \Psi\cdot{}_rj^m(f) \pitchfork (W\times\Delta) \quad \text{by (62)}\\ &\iff p_1\cdot\Psi\cdot{}_rj^m(f)\Big|_{e(X^{(r)}\times M)} \pitchfork W \quad \text{by (63)}\\ &\iff {}_r\tilde{j}^m(f) \pitchfork W \quad \text{by (61)} .\end{aligned}$$

More generally,

$$ {}_r j^m(f\Big|_{(X-C)\times M}) \pitchfork W' \iff {}_r \tilde{j}^m(f\Big|_{(X-C)\times M}) \pitchfork W $$

Now Lemma 32 applies. QED

An easy consequence of this is:

LEMMA 34: Let X, Y, C, M, f_0 be as in Lemma 33. Let $\{W_\alpha\}$ be a countable collection of submanifolds of various ${}_{r_\alpha}J^{m_\alpha}(X,Y)$. Then for every neighborhood N of f_0 in $C^\infty(X\times M, Y; f_0, C\times M)$ there exists a smooth homotopy f^u with

(64) $$ f^0 = f_0 $$

(65) $$ f^u \in N \quad \text{for all} \quad u $$

(66) $$ {}_{r_\alpha} j^{m_\alpha}(f^1\Big|_{(X-C)\times M}) \pitchfork W \quad \text{for all} \quad \alpha. $$

PROOF of Lemma 34: By Lemmas 33 and 31 there exists f^1 in N satisfying (66). Thus if N were "connected by smooth paths" the proof would be complete. But N contains a neighborhood N' of f_0 which is connected by smooth paths. QED

In our applications of Lemma 34, X will be $I\times P$, Y will be N, and $f\colon I\times P\times M \longrightarrow N$ will be one component of a fibered concordance

$$F = (h,f,p_3)\colon I\times P\times M \longrightarrow I\times N\times M .$$

C will be $I\times\partial P$.

§I.E. Approximation by Polynomials.

This section is concerned with the problem of approximating a (smooth or holomorphic) function f by polynomials g, in the sense of specifying that the function f-g should vanish at some finite set of points. Multiplicities are assigned to the points and f-g is required to vanish to higher order at multiple points. The main goal here is Lemma 38, which provides a useful link between smooth functions and polynomial functions. This is necessary in order to know that some algebraic manipulations of multijet sets in Ch.II (notably Operation D) will be relevant to the smooth category. (Lemma 61, §II.B.4). In effect what we are doing here is beginning to make the connection between the "Hilbert variety", which does not appear explicitly, and multijets, which do. (see Remark 4 in §D.3.b of the Introduction).

NOTATION: For any integers $k \geq 0$ and $m \geq 0$, denote by $Poly_k(\mathbb{R}^m)$ the space of all polynomial functions in $\mathbb{R}^m$ of degree less than or equal to k. Likewise let $Poly_k(\mathbb{C}^m)$ be the space of all complex polynomial functions in $\mathbb{C}^m$ of degree $\leq k$.

§I.E.1. The holomorphic case.

Consider first the case of functions of one variable. Fix an open subset $U \subset \mathbb{C}$ and let $H(U)$ be the space of all holomorphic functions from U to $\mathbb{C}$ (with the strong Whitney topology).

LEMMA 35. Let U be as above, let $k \geq 0$ be an integer, and let $a=(a_1,\cdots a_k) \in U^k$. Then for any $f \in H(U)$ there is a unique polynomial $A_{(a,f)} \in Poly_{k-1}(\mathbb{C})$ such that

(67) $f - A_{(a,f)}$ vanishes at $a_1,\cdots a_k$ "with multiplicities"

(i.e., if some complex number c appears N times in the list $a_1,\cdots a_k$ then $f-A_{(a,f)}$ vanishes to order $\geq N$ at c.) Moreover, $A_{(a,f)}$ depends continuously on (a,f).

PROOF of Lemma 35. Construct $A_{(a,f)}$ by induction on k. For $k=0$ set $A_{(a,f)}=0$. Then (67) holds vacuously. For the inductive step, let $k>0$ and $a=(a_1,\cdots a_k)\in U^k$. Note that the holomorphic function $f-f(a_k)$ vanishes at a_k, so that the function $\tilde{f}$ defined by

$$\tilde{f}(t) = \frac{f(t)-f(a_k)}{t-a_k} \qquad \text{for } t \neq a_k$$

$$\tilde{f}(a_k) = \frac{df}{dt}(a_k)$$

in U is holomorphic. Set

$$A_{(a_1,\cdots a_k,f)}(t) = f(a_k) + (t-a_k)A_{(a_1,\cdots a_{k-1},\tilde{f})}(t) \ .$$

Thus

$$f(t) - A_{(a_1,\cdots a_k,f)}(t) =$$

$$(t-a_k)(\tilde{f}(t) - A_{(a_1,\cdots a_{k-1},\tilde{f})}(t))$$

vanishes at $a_1, \cdots a_k$ "with multiplicities". This proves existence.

Uniqueness is clear; if there were two solutions, then their difference would be a nonzero polynomial of degree less than k with at least k roots.

The statement concerning continuous dependence is obvious from the construction. QED

The case of functions of more than one variable is not so simple; in particular solutions to interpolation problems tend to be non-unique. However, there is an easy existence statement (Lemma 36) and also a kind of continuity statement (Lemma 37).

LEMMA 36: Let $m \geq 0$, $r \geq 1$, and $k_i \geq 1$ $(1 \leq i \leq r)$ be integers. Set $K = \sum_{i=1}^{r} k_i$. Let $x=(x_1, \cdots x_r) \in (\mathbb{C}^m)^{(r)}$. Then the condition:

$$g \text{ vanishes to order } \geq k_i \text{ at } x_i \text{ for } 1 \leq i \leq r$$

defines a subspace of codimension $\sum_{i=1}^{r} \binom{m+k_i-1}{k_i-1}$ in $\mathrm{Poly}_{K-1}(\mathbb{C}^m)$.

PROOF of Lemma 36: This is equivalent to the statement that it is possible to find a polynomial g of degree $<K$ having any prescribed Taylor polynomials of degree k_j-1 at the x_j. In proving this statement it suffices to consider the case in which only one of the prescribed polynomials, say at x_i, is different from zero. Choose, for each $j \neq i$, a polynomial L_j of degree one such that

$$L_j(x_j) = 0, L_j(x_i) \neq 0$$

g will be taken to be a product

$$g = h \cdot \prod_{j \neq i} L_j^{k_j} ,$$

where h is a polynomial (as yet unchosen) of degree $<k_i$. To see that a suitable h exists, note that because

$$\prod_{j \neq i} L_j^{k_j}(x_i) \neq 0 ,$$

the linear endomorphism

$$h \longmapsto (k_i\text{-}1)\text{st Taylor polynomial of } g \text{ at } x_i$$

of $\mathrm{Poly}_{k_i-1}(\mathbb{C}^m)$ is invertible. QED

LEMMA 37: Let $m \geq 0$, $r \geq r' \geq 1$, and $k_i \geq 0 (1 \leq i \leq r)$ be integers. Let $K = \sum_{i=1}^{r} k_i$. Let

$$\{1, \cdots r\} \xrightarrow{\phi} \{1, \cdots r'\}$$

be surjective. Let $U \subset \mathbb{C}^m$ be open. Let f^ν be a convergent sequence in $H(U)$ with limit f. Let $x^\nu = (x_1^\nu, \cdots x_r^\nu)$ be a sequence in $U^{(r)}$ with limit $x = (x_1, \cdots x_r) = \phi^*(y) \underset{\text{def}}{=} (y_{\phi(1)}, \cdots y_{\phi(r)})$, where

$$y = (y_1, \cdots y_{r'}) \in U^{(r')}$$

Then there exists a convergent sequence g^ν in $\mathrm{Poly}_{K-1}(\mathbb{C}^m)$ with limit g such that

(68) $g^\nu - f^\nu$ vanishes to order k_i at x_i^ν for all ν, i

(69) $g - f$ vanishes to order $\sum_{\phi(i)=j} k_i$ at y_j for all j.

PROOF of Lemma 37: Use double induction on m and K. The cases $m=0$ and $K=0$ are trivial. The case $m=1$ follows easily from Lemma 35 by setting:

$$a^\nu = (\underbrace{x_1^\nu, \cdots x_1^\nu}_{k_1 \text{ times}}, \cdots \underbrace{x_r^\nu, \cdots x_r^\nu}_{k_r \text{ times}}) \quad \text{for all} \quad \nu \geq 1$$

$$a = (\overbrace{x_1, \cdots x_1}, \cdots \overbrace{x_r, \cdots x_r})$$

$$g^\nu = A_{(a^\nu, f^\nu)}$$

$$g = A_{(a,f)}$$

Thus g^ν converges to g and (68) and (69) are satisfied.

Now let $m \geq 2$ and $K \geq 1$, and assume that Lemma 37 is true for $(m-1,K)$ and for $(m,K-1)$. Since $K \geq 1$, one of the $\{k_i\}$ is positive. Without loss of generality $k_r \geq 1$.

<u>Step 1</u>. The conclusion holds if one assumes the additional hypotheses:

(70) $$x_r^\nu = 0 \quad \text{for all} \quad \nu$$

(71) $$x_i = 0 \quad \text{for all} \quad i .$$

(Thus $r' = 1$ and (69) becomes

(72) $$g - f \text{ vanishes to order } K \text{ at } 0 .)$$

Without loss of generality (by a linear automorphism of $\mathbb{C}^m$) it can be assumed that the projection

$$p_1 : \mathbb{C}^m = \mathbb{C}^{m-1} \times \mathbb{C} \longrightarrow \mathbb{C}^{m-1}$$

satisfies

(73) $$p_1(x_i^\nu) \neq p_1(x_j^\nu) \quad \text{for all } \nu, \text{ if } i \neq j .$$

(Here the assumption $m \geq 2$ is needed.) Choose open neighborhoods U_1 of 0 in $\mathbb{C}^{m-1}$ and U_2 of 0 in $\mathbb{C}$, such that $U_1 \times U_2 \subset U$. Except for finitely many indices ν, then, $x_i^\nu \in U_1 \times U_2$ for all i, by (71). But any finite set of values ν can safely be ignored, because by Lemma 36 one can find $g^\nu \in \mathrm{Poly}_{K-1}(\mathbb{C}^m)$ for these missing ν's such that (68) holds. Define f_1^ν and f_1 in $H(U_1)$ and f_2^ν and f_2 in $H(U_1 \times U_2)$ by writing

$$f^\nu(x,t) = f_1^\nu(x) + t f_2^\nu(x,t)$$

$$f(x,t) = f_1(x) + t f_2(x,t)$$

for any $(x,t) \in U_1 \times U_2$. Then the hypotheses of Lemma 37 are satisfied with

$$(m-1,r,\{k_i\},K,U_1,\{f_1^\nu\},f_1,\{p_1(x_i^\nu)\},\{p_1(x_i)\})$$

in place of

$$(m,r,\{k_i\},K,U,\{f^\nu\},f,\{x_i^\nu\},\{x_i\}) .$$

Thus by induction on m there exist $g_1^\nu \in \mathrm{Poly}_{K-1}(\mathbb{C}^{m-1})$ converging to some $g_1 \in \mathrm{Poly}_{K-1}(\mathbb{C}^{m-1})$ such that

(74) $g_1^\nu - f_1^\nu$ vanishes to order k_i at $p_1(x_i^\nu)$ for all ν, i

(75) $g_1 - f_1$ vanishes to order K at 0 .

On the other hand, the hypotheses are also satisfied by

$$(m,r,(\{k_i\}_{i=1}^{r-1},k_r-1),K-1,U_1 \times U_2,\{f_2^\nu\},f_2,\{x_i^\nu\},\{x_i\}) .$$

Thus by induction on K there exist $g_2^\nu \in \mathrm{Poly}_{K-2}(\mathbb{C}^m)$ converging to $g_2 \in \mathrm{Poly}_{K-2}(\mathbb{C}^m)$ such that

$$(76)\quad g_2^\nu - f_2^\nu \text{ vanishes to order } \begin{cases} k_i & \text{at } x_i^\nu \text{ for } i \neq r \\ k_r - 1 & \text{at } 0 \end{cases}$$

$$(77)\quad g_2 - f_2 \quad \text{vanishes to order} \quad K-1 \quad \text{at} \quad 0 .$$

Set

$$g^\nu(x,t) = g_1^\nu(x) + tg_2^\nu(x,t)$$

$$g(x,t) = g_1(x) + tg_2(x,t) .$$

Then (68) follows from (74) and (76), while (72) follows from (75) and (77).

Step 2. The conclusion holds if one assumes (71) but not (70). To see this, choose $\epsilon > 0$ such that

$$\overline{D_{2\epsilon}(0)} \subset U$$

(Here $D_{2\epsilon}(0)$ is the open disk of center 0 and radius 2ϵ.) Then except for finitely many ν, which can be ignored as in Step 1,

$$x_i^\nu \in D_\epsilon(0) \quad \text{for all} \quad \nu \quad \text{and} \quad i .$$

Now set

$$(78)\qquad \begin{cases} \tilde{U} = D_\epsilon(0) \\ \tilde{f}^\nu(\xi) = f\,(\xi + x_r^\nu) \quad \text{for} \quad \xi \in \tilde{U} \\ \tilde{f} = f\Big|_{\tilde{U}} \\ \tilde{x}_i^\nu = x_i^\nu - x_r^\nu \\ \tilde{x}_i = 0 \end{cases}$$

Substituting $\tilde{U}$ for U, $\tilde{f}^\nu$ for f^ν, $\tilde{f}$ for f, and $\tilde{x}_i^\nu$ for x_i^ν, one has the hypotheses of Lemma 37 and (70) and (71). Therefore Step 1 applies, so that there exist $\tilde{g}^\nu$ converging to some $\tilde{g}$ in $\mathrm{Poly}_{K-1}(\mathbb{C}^m)$ such that

(79) $\tilde{g}^\nu - \tilde{f}^\nu$ vanishes to order k_i at x_i^ν for all ν, i

(80) $\tilde{g} - \tilde{f}$ vanishes to order K at 0 .

Set

$$(81)\qquad \begin{cases} g^\nu(\xi) = \tilde{g}^\nu(\xi - x_r^\nu) \\ g = \tilde{g} \end{cases} .$$

Then (79) and (80) imply (68) and (72), hence (69), using (78) and (81).

Step 3: Assume only that x_i is independent of i (in other words that $r'=1$), but do not assume (71). The conclusion follows in this case by using a translation in $\mathbb{C}^m$ to get back to Step 2.

Step 4: The general case: Fix j $(1 \leq j \leq r')$. Polynomials g_j^ν and g_j in $\mathrm{Poly}_{K-1}(C^m)$ will be found such that

(82) $\lim\limits_{\nu\to\infty} g_j^\nu = g_j$

(83) $g_j^\nu - f^\nu$ vanishes to order k_i at x_i^ν if $\phi(i)=j$

(84) g_j^ν vanishes to order k_i at x_i^ν if $\phi(i)\neq j$

(85) $g_j - f$ vanishes to order $\sum\limits_{\phi(i)=j} k_i$ at y_j

(86) g_j vanishes to order $\sum\limits_{\phi(i)=j'} k_i$ at $y_{j'}$, for $j'\neq j$

This will suffice, for then if one sets

$$g^\nu = \sum_{j=1}^{r'} g_j^\nu$$
$$g = \sum_{j=1}^{r'} g_j$$

then g^ν and g will satisfy (68) and (69).

To find g_j^ν and g_j, first choose a linear polynomial $L \in \mathrm{Poly}_1(\mathbb{C}^m)$ such that

$$L(x_i^\nu) \neq L(x_{i'}^\nu) \quad \text{when} \quad i \neq i'$$

$$L(y_j) \neq L(y_{j'}) \quad \text{when} \quad j \neq j' \quad .$$

Then set

$$q_j^\nu = \prod_{\phi(i)\neq j} (L - L(x_i^\nu))^{k_i}$$

$$q_j = \prod_{\phi(i)\neq j} (L - L(y_i))^{k_i} \quad .$$

Clearly q_j^ν converges to q_j in $\mathrm{Poly}_{K-\sum_{\phi(i)=j} k_i}(\mathbb{C}^m)$, and

(87) $\quad q_j^\nu(x_i^\nu) \neq 0$ if $\phi(i) = j$

(88) $\quad q_j^\nu$ vanishes to order k_i at x_i^ν if $\phi(i) \neq j$

(89) $\quad q_j(y_j) \neq 0$

(90) q_j vanishes to order $\sum_{\phi(i)=j'} k_i$ at $y_{j'}$, if $j' \neq j$.

Now polynomials h_j^ν and h_j in $\mathrm{Poly}_{\sum_{\phi(i)=j} k_i - 1}(\mathbb{C}^m)$ will be found, such that

(91) $\lim h_j^\nu = h_j$

(92) $h_j^\nu - \dfrac{f^\nu}{q_j^\nu}$ vanishes to order k_i at x_i^ν if $\phi(i)=j$

(93) $h_j - \dfrac{f}{q_j}$ vanishes to order $\sum_{\phi(i)=j} k_i$ at y_j .

(Notice that (92) and (93) make sense, by (87) and (89).) To find h_j^ν and h_j use Step 3, numbering all those i such that $\phi(i)=j$ in some way:

$$i_1, \cdots i_{\hat{r}}$$

and applying Step 3 to the data:

$$(m, \hat{r}, \left\{k_{i_\alpha}\right\}_{\alpha=1}^{\hat{r}}, \sum_{\alpha=1}^{\hat{r}} k_{i_\alpha}, \hat{U}, \hat{f}^\nu, \hat{f}, \left\{\hat{x}_\alpha^\nu\right\}, \left\{\hat{x}_\alpha\right\}) \quad ,$$

where

$$\hat{U} = \text{interior of } \left\{ \xi \epsilon U \,\middle|\, q_j(\xi) \neq 0 \text{ and for all } \nu\, q_j^\nu(\xi) \neq 0 \right\}$$

$$\hat{f}^\nu = \frac{f^\nu}{q_j^\nu} \quad \text{in} \quad \hat{U}$$

$$\hat{f} = \frac{f}{q_j} \quad \text{in} \quad \hat{U}$$

$$\hat{x}_\alpha^\nu = x_{i_\alpha}^\nu$$

$$\hat{x}_\alpha = x_{i_\alpha} \quad (=y_j) \ .$$

Finally, given $\left\{h_j^\nu\right\}$ and f_j satisfying (92) and (93), define

$$g_j^\nu = h_j^\nu \, q_j^\nu$$

$$g_j = h_j \, q_j \ .$$

(82)-(86) are easily checked, using (88), (90), (91), (92), and (93). QED

§I.E.2. The C^∞ case.

Finally, it is easy to check that all of §I.E.1 remains true if U is open in $\mathbb{R}^m$ rather than $\mathbb{C}^m$ and $H(U)$ is replaced by the space $C^\infty(U)$ of smooth functions from U to $\mathbb{R}$ and $\mathrm{Poly}_{K-1}(\mathbb{C}^m)$ is replaced by $\mathrm{Poly}_{K-1}(\mathbb{R}^m)$. In particular we have

LEMMA 38. Let $m \geq 0$, $r \geq r' \geq 1$, and $k_i \geq 0$ $(1 \leq i \leq r)$ be integers. Let $K = \sum_{i=1}^{r} k_i$. Let

$$\{1, \cdots r\} \xrightarrow{\phi} \{1, \cdots r'\}$$

be surjective. Let $U \subset \mathbb{R}^m$ be open. Let f^ν be a convergent sequence in $C^\infty(U)$ with limit f. Let $x^\nu = (x_1^\nu, \cdots x_r^\nu)$ be a sequence in $U^{(r)}$ with limit $x = (x_1, \cdots x_r) = \phi^*(y) \underset{\text{def}}{=} (y_{\phi(1)}, \cdots y_{\phi(r)})$, where

$$y = (y_1, \cdots y_{r'}) \in U^{(r')} .$$

Then there exists a convergent sequence g^ν in $\mathrm{Poly}_{K-1}(\mathbb{R}^m)$ with limit g such that

(94) $g^\nu = f^\nu$ vanishes to order k_i at x_i^ν for all ν, i.

(95) $g - f$ vanishes to order $\sum_{\phi(i)=j} k_i$ at y_j for all j.

Chapter II. The Collection Z of Multijet Sets.

In proving Theorem D in the next chapter we will make extensive use of certain subsets $p_m^{\infty}S(Z,P,N)$ of the multijet manifolds ${}_rJ^m(\mathbb{R}\times P,N)$ associated with manifolds P^p and N^n. The purpose of Chapter II is to define and investigate these sets.

The sets will not be manifolds. Their saving grace is that they are defined, in terms of local coordinates in P and N, by polynomial equations (in a way which is independent of the choice of coordinates). Thus the sets correspond to certain algebraic varieties in ${}_rJ^m(\mathbb{R}\times\mathbb{R}^p, \mathbb{R}^n)$.

The sets which are forced upon us by the geometry are slightly smaller than the sets $p_m^{\infty}S(Z,P,N)$ which we will actually use. They correspond to certain semi-algebraic subsets of ${}_rJ^m(\mathbb{R}\times\mathbb{R}^p,\mathbb{R}^n)$ - sets defined by polynomial equations and polynomial inequalities. The main point here is that a semi-algebraic variety has the same dimension as the smallest algebraic variety which contains it.

Since the proof of Theorem D is fundamentally a (very complicated) general position argument, showing that for dimensional reasons certain things can be made disjoint, it is not too surprising that the larger algebraic varieties should serve as well as the smaller semi-algebraic ones.

In any case we completely avoid semi-algebraic sets, and as far as possible we even avoid real algebraic sets, preferring to do our algebraic geometry over the complex numbers (for reasons like Fact 7 of Ch.I. §A.2, which is false in the real case). Thus the dimensions which are actually calculated or estimated are complex dimensions of complex algebraic varieties in the space of complex multijets ${}_rJ^m_{\mathbb{C}}(\mathbb{C}\times\mathbb{C}^p,\mathbb{C}^n)$.

The plan of Chapter II is this:

§A. Define and study the notion of "invariant algebraic set of complex multijets" (IASCM). "Invariant" here refers to the independence of the set on the choice of coordinates in $\mathbb{C}^p$ and $\mathbb{C}^n$.

§B. Define operations A,B,C, and D (all forced on us by the proof of Theorem D) which make new IASCM's out of old.

§C. Use the operations A,B,C, and D to define the collection $\mathcal{Z}$ of IASCM's which will be used in Chapter III.

§D. Investigate the subsets $S(Z,P,N)$ which are determined by elements $Z \in \mathcal{Z}$ and manifolds P and N, and begin to see how these behave with respect to fibered concordances of P in N.

Perhaps a few words of explanation are in order here because of the occurrence of holomorphic mappings in the proofs of same lemmas. It would have been simpler and clearer if all of the complex geometry could have been kept algebraic. However, at one point we saw no way out but to integrate a polynomial vector field. (This is in the proof of Lemma 77 (ix).) The upshot is that we will see three different versions of "invariance of multijet sets under coordinate change":

(96) Real multijet sets and C^{∞} coordinate changes (the version we are originally interested in)

(97) Complex multijet sets and coordinate changes on the jet level (the version where it is easiest to study the multijet sets as algebraically defined objects), and

(98) Complex multijet sets and holomorphic coordinate changes (which is introduced as an aid in studying (97)).

§II.A. Invariant Algebraic Sets of Complex Multijets.

§II.A.1. Algebraic Sets of Complex Multijets.

Define (for $1 \leq r < \infty$ and $0 \leq m \leq \infty$) the set of complex multijets ${}_rJ^m_{\mathbb{C}}(\mathbb{C}^{p+1},\mathbb{C}^n)$ by mimicking the definition (§I.D.) of real multijets. (Use local holomorphic maps from $\mathbb{C}^{p+1}$ to $\mathbb{C}^n$ instead of smooth maps, and use complex Taylor polynomials instead of real ones.) For short, write

$$ {}_rJ^m = {}_rJ^m_{\mathbb{C}}(\mathbb{C}^{p+1},\mathbb{C}^n) $$

Define maps

$$ {}_rJ^m \xrightarrow{\ s\ } (\mathbb{C}^{p+1})^{(r)} \qquad \text{for} \quad 0 \le m \le \infty $$

$$ {}_rJ^m \xrightarrow{\ t\ } (\mathbb{C}^n)^r \qquad \text{for} \quad 0 \le m \le \infty $$

$$ {}_rJ^{m'} \xrightarrow{\ p^{m'}_m\ } {}_rJ^m \qquad \text{for} \quad 0 \le m \le m' \le \infty $$

as in the real case. It is easy to see that for $m<\infty$ ${}_rJ^m$ has the structure of a smooth complex algebraic variety, in fact a (Zariski) open set in an affine space. Moreover, ${}_rJ^m$ has a real structure. (That is, it is defined by real polynomials.)

DEFINITION 39: Let $m \geq 0$ and $r \geq 1$ be integers. A subset $Z \subset {}_rJ^\infty$ is called algebraic of level m and rank r if

$$Z = (p_m^\infty)^{-1}(p_m^\infty(Z))$$

and

$p_m^\infty(Z)$ is a closed algebraic subset of ${}_rJ^m$ defined by real polynomials.

Note that if Z has level m then it also has level m' for all $m' \geq m$.

EXAMPLE 40: Let

$$Z^0 = \left\{ (z_1, z_2) \in {}_2J^\infty \;\middle|\; t(z_1) = t(z_2) \right\}.$$

Z^0 is clearly an algebraic set of rank 2 and level 0. This example corresponds to the set S_0 in the Introduction (Intro.D.2). It will play a fundamental role in the proof of Theorem D.

§II.A.2. Invariance.

If an algebraic set Z of rank r is suitably "invariant under coordinate change" (Def. 45 below), then it can be used to construct sets

$$S(Z,P,N) \subset {}_rJ^\infty(\mathbb{R}\times P,N)$$

for every pair of smooth manifolds P and N with respective dimensions p and n. Before making the definition of "invariance" let us see how the construction of $S(Z,P,N)$ will go.

${}_rJ^\infty(\mathbb{R}\times P,N)$ can be covered by sets of the form ${}_rJ^\infty(\mathbb{R}\times U,V)$, where U and V are open subsets of P and N respectively and admit embeddings*

---

*Actually this last statement is false if P is a 0-dimensional manifold consisting of more than one point. However, we can get around this as follows:

Suppose Theorem D is known in the case when P is a point. Then it can be proved in the case when P is a finite set, by induction on the cardinality of P. Write P as the disjoint union of nonempty P_1 and P_2. There are fibrations

$$C(P_1,N-P_2) \to C(P,N) \to C(P_1,N)$$

and more generally for $S \subset \{1,\cdots a\}$

$$C(P_1,N-P_2-\bigcup_{j\in S} Q_j) \to C(P,N-\bigcup_{j\in S} Q_j) \to C(P_1,N-\bigcup_{j\in S} Q_j) .$$

These give rise to a long exact sequence of homotopy groups of $(a+1)$-ads. By induction the conclusion of Theorem D holds for two of the $(a+1)$-ads, and therefore it also holds for the third.

Alternatively Theorem D can be proved directly in the case $p=0$ by a rather easy general-position argument.

$$U \xrightarrow{\ \phi\ } \mathbb{R}^p$$

$$V \xrightarrow{\ \psi\ } \mathbb{R}^n \quad .$$

The idea is then this: Identify ${}_rJ^\infty(\mathbb{R}^{p+1},\mathbb{R}^n)$ with a subset of ${}_rJ^\infty$ in the obvious way, and for $z \in {}_rJ^\infty(\mathbb{R}\times U,V)$ define

$$(99) \qquad z \in S(Z,P,N) \iff \text{for some } \tilde{z} \in Z$$

$$ {}_rj^\infty(\psi)(tz)\cdot z = \tilde{z}\cdot {}_rj^\infty(1_{\mathbb{R}}\times\phi)(sz)$$

Of course, for (99) to make sense it is necessary for the condition on the right-hand side of (99) to be independent of the choice of U and V such that $z \in {}_rJ^\infty(\mathbb{R}\times U,V)$, and also independent of the choice of ϕ and ψ. This is part of the reason for the definition (Def 45) of "invariant" below.

The results of the next two subsections II.A.2.a and II.A.2.b simply serve to link together the three kinds of coordinate change ((96) (97) and (98)) mentioned in the introduction to Ch.II. We shall deal separately with changes of coordinate in the domain $(\mathbb{C}^{p+1})$ and in the range $(\mathbb{C}^n)$.

§II.A.2.a. Domain-Invariance.

The coordinate changes in the domain will have the form (U,V), where U and V are complex diffeomorphisms between open subsets of $\mathbb{C}^{p+1}$ and of $\mathbb{C}^p$ respectively, fitting into a commutative diagram

$$\begin{array}{ccccccc} \mathbb{C}^{p+1} & \supset & \Omega_1 & \xrightarrow{U} & \Omega_2 & \subset & \mathbb{C}^{p+1} \\ \downarrow{\scriptstyle p_2} & & \downarrow & & \downarrow & & \downarrow{\scriptstyle p_2} \\ \mathbb{C}^p & \supset & \Omega_3 & \xrightarrow{V} & \Omega_4 & \subset & \mathbb{C}^p \end{array} \quad . \tag{100}$$

(Here p_2 is the projection of $\mathbb{C}^{p+1} = \mathbb{C}\times\mathbb{C}^p$ onto the second factor.) Given such U and V, define for each $m \geq 1$ and $r \geq 1$ a map

$${}_rJ_{\mathbb{C}}^m(\Omega_2,\mathbb{C}^n) \xrightarrow{U^*} {}_rJ_{\mathbb{C}}^m(\Omega_1,\mathbb{C}^n)$$

by

$$U^*(z)_i = z_i \cdot {}_rj^m(U)(U^{-1}sz_i), \quad 1\leq i\leq r \ . \tag{101}$$

$\mathcal{U}^*$ is a complex diffeomorphism (with inverse $(\mathcal{U}^{-1})^*$). An algebraic set $Z \subset {}_rJ^\infty$ of level $m \geq 1$ will be called <u>domain</u>-<u>invariant</u> if, for every diagram of the form (100), we have

$$(102)\qquad \begin{cases} \mathcal{U}^*(p_m^\infty(Z) \cap {}_rJ_{\mathbb{C}}^m(\Omega_2, \mathbb{C}^n)) \\ \quad = p_m^\infty(Z) \cap {}_rJ_{\mathbb{C}}^m(\Omega_1, \mathbb{C}^n) \ . \end{cases}$$

Notice that for any $(x_1, \cdots x_r) \in \Omega_1{}^{(r)}$ the jets

$$u_i \underset{\text{def}}{=} j^\infty(\mathcal{U})(x_i) \in J_{\mathbb{C}}^\infty(\Omega_1, \Omega_2) \subset J_{\mathbb{C}}^\infty(\mathbb{C}^{p+1}, \mathbb{C}^{p+1})$$

$$v_i \underset{\text{def}}{=} j^\infty(\mathcal{V})(p_2 x_i) \in J_{\mathbb{C}}^\infty(\Omega_3, \Omega_4) \subset J_{\mathbb{C}}^\infty(\mathbb{C}^p, \mathbb{C}^p)$$

satisfy

(103) $\qquad u_i$ and v_i are invertible

(104) $\qquad su_i \neq su_j$ if $i \neq j$

(105) $\qquad tu_i \neq tu_j$ if $i \neq j$

(106) $\quad j^\infty(p_2)(tu_i) \cdot u_i = v_i \cdot j^\infty(p_2)(su_i)$ for all i.

(107) $sv_i = sv_j \Longleftrightarrow tv_i = tv_j \Longleftrightarrow v_i = v_j$ for all i and j.

LEMMA 41: An algebraic set $Z \subset {}_rJ^\infty$ is domain-invariant if and only if for all $u_1,\cdots u_r \in J^\infty_{\mathbb{C}}(\mathbb{C}^{p+1},\mathbb{C}^{p+1})$ and $v_1,\cdots v_r \in J^\infty_{\mathbb{C}}(\mathbb{C}^p,\mathbb{C}^p)$ satisfying (103)-(107) and for all $z \in Z$ such that

(108) $$sz_i = tu_i \quad \text{for} \quad 1 \le i \le r$$

the multijet $z\cdot u = (z_1\cdot u_1,\cdots z_r\cdot u_r)$ is in Z.

PROOF of Lemma 41: The "if" statement follows from the discussion preceding the Lemma. For the "only if" it is necessary to reconstruct U and V from u and v. More precisely, suppose that Z is algebraic of level $m \ge 1$, $z \in Z$, and (103)-(108) hold. The plan is to find polynomial mappings

$$\mathbb{C}^{p+1} \xrightarrow{\ U\ } \mathbb{C}^{p+1}$$

$$\mathbb{C}^p \xrightarrow{\ V\ } \mathbb{C}^p$$

such that

(109) $$p_2 U = V p_2$$

and such that for all i $(1 \le i \le r)$

(110) $$j^m(U)(su_i) = p_m^\infty(u_i)$$

(111) $$j^m(V)(sv_i) = p_m^\infty(v_i)$$

This will imply that for suitable open sets $\Omega_1 \subset \mathbb{C}^{p+1}$ containing $\{su_i, \cdots su_r\}$ and $\Omega_3 \subset \mathbb{C}^p$ containing $\{sv_1, \cdots sv_r\}$

(112) $$\begin{cases} U|_{\Omega_1} & \text{is an embedding} \\ V|_{\Omega_3} & \text{is an embedding} \\ p_2(\Omega_1) \subset \Omega_3 & . \end{cases}$$

Indeed, (111) and (103) imply that V embeds some neighborhood of each sv_i. Then (111) and (107) imply that $V(sv_i) \neq V(sv_j)$ when $sv_i \neq sv_j$. Therefore V embeds some neighborhood Ω_3 of $\{sv_1,\cdots sv_r\}$. Likewise (using (110) in place of (111) and (105) in place of (107) $\mathcal{U}$ embeds some neighborhood of $\{su_1,\cdots su_r\}$ which can be taken to be contained in $p_2^{-1}(\Omega_3)$.

Now to find $\mathcal{U}$ and $\mathcal{V}$ satisfying (109)-(111). First of all, by (107) there is certainly a polynomial mapping

$$\mathbb{C}^p \xrightarrow{\mathcal{V}} \mathbb{C}^p$$

satisfying (111). Also, by (104) there is a polynomial function

$$\mathbb{C}^{p+1} \xrightarrow{\mathcal{U}_1} \mathbb{C}$$

satisfying

$$(113) \qquad j^m(\mathcal{U}_1)(su_i) = j^m(p_1)(tu_i)\cdot p_m^\infty(u_i)$$

for all i. Set

$$\mathcal{U} = (\mathcal{U}_1, \mathcal{V}\cdot p_2) \ .$$

Then (109) is obvious and (110) follows from

$$\begin{aligned} j^m(p_1)(tu_i).j^m(U)(su_i) &= j^m(p_1\cdot U)(su_i) \\ &= j^m(U_1)(su_i) \\ &= j^m(p_1)(tu_i)\cdot p_m^\infty(u_i) \end{aligned}$$

and

$$\begin{aligned} j^m(p_2)(tu_i)\cdot j^m(U)(su_i) &= j^m(p_2U)(su_i) \\ &= j^m(Vp_2)(su_i) \\ &= j^m(V)(p_2su_i)\cdot j^m(p_2)(su_i) \\ &= p_m^\infty(v_i)\cdot p_m^\infty(j^\infty(p_2)(su_i)) \\ &= p_m^\infty(v_i\cdot j^\infty(p_2)(su_i)) \\ &= p_m^\infty(j^\infty(p_2)(tu_i)\cdot u_i) \\ &= j^m(p_2)(tu_i)\cdot p_m^\infty(u_i)\ . \end{aligned}$$

Now (112) implies that this is an instance of (100). Therefore the definition of domain-invariance applies:

$$p_m^\infty(z_i \cdot u_i) = p_m^\infty(z_i) \cdot p_m^\infty(u_i)$$

$$= p_m^\infty(z_i) \cdot j^m(\mathcal{U})(su_i) \quad \text{by (110)}$$

$$= \mathcal{U}^*(p_m^\infty(z))_i \ .$$

Therefore $p_m^\infty(z \cdot u) = \mathcal{U}^*(p_m^\infty(z)) \in p_m^\infty(Z)$, by domain-invariance of Z, and $z \cdot u \in Z$. QED

As a corollary of the proof of Lemma 41 one can prove the following C^∞ result:

LEMMA 42: Let Z be algebraic of rank r and level m and domain-invariant. Then for any (C^∞) diffeomorphisms $\mathcal{U}$ and $\mathcal{V}$ between open sets in $\mathbb{R}^{p+1}$ and $\mathbb{R}^p$, fitting into a commutative diagram

$$\begin{array}{ccccccc} \mathbb{R}^{p+1} & \supset & \Omega_1 & \xrightarrow{U} & \Omega_2 & \subset & \mathbb{R}^{p+1} \\ p_2 \downarrow & & \downarrow & & \downarrow & & \downarrow p_2 \\ \mathbb{R}^{p} & \supset & \Omega_3 & \xrightarrow{V} & \Omega_4 & \subset & \mathbb{R}^{p} \end{array}$$

we have

$$U^*(p_m^\infty Z \cap {}_rJ^m(\Omega_2, \mathbb{R}^n)) = p_m^\infty Z \cap {}_rJ^m(\Omega_1, \mathbb{R}^n) \ .$$

(Here $U^*: {}_rJ^m(\Omega_2, \mathbb{R}^n) \longrightarrow {}_rJ^m(\Omega_1, \mathbb{R}^n)$ is defined by (101) as in the complex case.)

§II.A.2.b. Range-Invariance.

The changes of coordinate in the range are simpler: they are just complex diffeomorphisms

$$\mathbb{C}^n \supset \Omega_1 \xrightarrow{W} \Omega_2 \subset \mathbb{C}^n$$

between open subsets of $\mathbb{C}^n$. Such a map W determines, for each $m \geq 1$ and $r \geq 1$, a complex diffeomorphism

$${}_rJ^m_{\mathbb{C}}(\mathbb{C}^{p+1}, \Omega_1) \xrightarrow{W^{**}} {}_rJ^m_{\mathbb{C}}(\mathbb{C}^{p+1}, \Omega_2) \ ,$$

namely

$$\mathcal{W}^{**}(z)_i = j^m(\mathcal{W})(tz_i)\cdot z_i \ .$$

An algebraic set Z of level m and rank r is called <u>range-invariant</u> if for all such $\mathcal{W}$ we have

$$\mathcal{W}^{**}(p_m^\infty(Z) \cap {}_rJ_{\mathbb{C}}(\mathbb{C}^{p+1},\Omega_1)) =$$

$$p_m^\infty(Z) \cap {}_rJ_{\mathbb{C}}^m(\mathbb{C}^{p+1},\Omega_2) \ .$$

Notice that for all such $\mathcal{W}$ and any $(y_1,\cdots y_r) \in \Omega_1^r$ the jets

$$w_i \underset{\text{def}}{=} j^\infty(\mathcal{W})(y_i) \in J_{\mathbb{C}}^\infty(\mathbb{C}^n,\Omega_1) \subset J_{\mathbb{C}}^\infty(\mathbb{C}^n,\mathbb{C}^n)$$

satisfy

(114) w_i is invertible for all i

(115) $sw_i = sw_j \iff tw_i = tw_j \iff w_i = w_j$, for all i and j.

Corresponding to Lemma 41 we have:

LEMMA 43: An algebraic set $Z \subset {}_rJ^\infty$ is range-invariant if and only if for all $w_1, \cdots w_r$ in $J^\infty_{\mathbb{C}}(\mathbb{C}^n, \mathbb{C}^n)$ satisfying (114) and (115) and for all $z \in Z$ such that

(116) $$tz_i = sw_i \quad \text{for all } i$$

the multijet

$$w \cdot z = (w_1 \cdot z_1, \cdots w_r \cdot z_r)$$

is in Z.

PROOF of Lemma 43: This resembles the proof of Lemma 41 but is simpler. Again the "if" part has already been proved. For the "only if", reconstruct $\mathcal{W}$ from $(w_1, \cdots w_r)$. By (115) there is a polynomial mapping

$$\mathbb{C}^n \xrightarrow{\mathcal{W}} \mathbb{C}^n$$

satisfying

$$j^m(\mathcal{W})(sw_i) = p^\infty_m w_i \quad \text{for all } i .$$

By (114) $\mathcal{W}$ embeds a neighborhood of each sw_i. By (115) $\mathcal{W}$ embeds a neighborhood Ω_1 of the set $\{sw_1, \cdots sw_r\}$. Let $\Omega_2 = \mathcal{W}(\Omega_1)$, and observe that

$$\begin{aligned} p_m^\infty(w_i \cdot z_i) &= p_m^\infty(w_i) \cdot p_m^\infty(z_i) \\ &= j^m(\mathcal{W})(sw_i) \cdot p_m^\infty(z_i) \\ &= (\mathcal{W}^{**}(p_m^\infty z))_i \quad . \end{aligned}$$

So

$$\begin{aligned} p_m^\infty(w \cdot z) &= \mathcal{W}^{**}(p_m^\infty z) \\ &\in p_m^\infty Z \end{aligned}$$

by range-invariance of Z, and since Z has level m $w \cdot z \in Z$. QED

Corresponding to Lemma 42 we have:

LEMMA 44. Let Z be algebraic of rank r and level m, and range-invariant. Then for any diffeomorphism

$$\mathbb{R}^n \supset \Omega_1 \xrightarrow{\omega} \Omega_2 \subset \mathbb{R}^n$$

between open sets in $\mathbb{R}^n$ we have

$$\omega^{**}(p_m^\infty Z \cap {}_rJ^m(\mathbb{R}^{p+1}, \Omega_1)) =$$

$$p_m^\infty Z \cap {}_rJ^m(\mathbb{R}^{p+1}, \Omega_2) \quad .$$

(Again, ω^{**} is defined as in the complex case.)

§II.A.2.c. S(Z,P,N).

DEF. 45: An invariant algebraic set of complex multi-jets (IASCM) is an algebraic set of complex multijets which is both domain-invariant and range-invariant.

The algebraic sets of multijets that will occur in this paper will all be IASCM's.

It is now easy to check that if Z is an IASCM then (99) makes sense: To check that the RHS of (99) is independent of ϕ use Lemma 42 (in the special case in which

$$\Omega_1 = \mathbb{R} \times \Omega_3$$

$$\Omega_2 = \mathbb{R} \times \Omega_4$$

$$\mathcal{U} = 1 \times \mathcal{V}) .$$

To check that the RHS of (99) is independent of ψ use Lemma 44. (The full strength of Lemma 42 will not be used until Lemma 53.) Now the following definition makes sense:

DEF. 46. If Z is an invariant algebraic set of rank r and level m, then for any smooth manifolds P^p and N^n the set $S(Z,P,N)$ is defined by (99).

EXAMPLE 47. The set Z^0 (see Example 40, §II.A.1) is easily seen to be an IASCM. Therefore for any P and N the set $S(Z^0,P,N)$ is defined. It is precisely

$$\{ (z_1,z_2) \in {}_2J^\infty(\mathbb{R} \times P,N) | tz_1 = tz_2 \}$$

The following is easy:

LEMMA 48. If Z is an IASCM of level m, then:

$$S(Z,P,N) = (p_m^\infty)^{-1} p_m^\infty(S(Z,P,N))$$

$$p_m^\infty(S(Z,P,N)) \text{ is a closed subset of } {}_rJ^m(\mathbb{R}\times P,N)$$

It was not assumed in Def. 39 or Def. 45 that $p_m^\infty(Z)$ was a nonsingular variety. However, any complex algebraic variety has a natural stratification into nonsingular varieties, and this will give rise to a stratification of $S(Z,P,N)$ into smooth manifolds. The main interest will be in the "top stratum" $S^*(Z,P,N) \subset S(Z,P,N)$, to be defined below.

DEFINITION 49. Let Z be an IASCM of level m and rank r. Then $A(Z) = (p_m^\infty)^{-1}(\Sigma)$, where Σ is the singular set of the complex algebraic variety $p_m^\infty(Z)$.

(For the definition of "singular set" see Remark 8 of §I.A.2.)

EXAMPLE 50: $A(Z^0) = \emptyset$.

It is clear (see Remark 8) that Σ is a closed algebraic subset of $p_m^\infty(Z)$. Therefore $A(Z)$ is an algebraic set of level m and rank r. It will be proved (Lemma 64) that $A(Z)$ is an IASCM if Z is. Thus for any P and N there is a subset

$$S(A(Z),P,N) \subset S(Z,P,N)$$

DEFINITION 51: Let Z be an IASCM and P^p and N^n smooth manifolds. Then

$$S^*(Z,P,N) = S(Z,P,N) - S(A(Z),P,N) .$$

It is easy to prove the following:

LEMMA 52: $S^*(Z,P,N) = (p_m^\infty)^{-1}(p_m^\infty(S^*(Z,P,N)))$, where m is the level of Z. $p_m^\infty(S^*(Z,P,N))$ is a smooth submanifold of ${}_rJ^m(\mathbb{R}\times P,N)$.

We also have:

LEMMA 53: Let Z be an IASCM of rank r and level m, and let P^p and N^n be manifolds. If

$$\mathbb{R}\times P\times D^s \xrightarrow{\;\;U\;\;} \mathbb{R}\times P\times D^s$$

is a diffeomorphism fibered over $P\times D^s$, then the diffeomorphism

$$_rJ^m(\mathbb{R}\times P,N)\times D^s \xrightarrow{\;\;U'\;\;} {}_rJ^m(\mathbb{R}\times P,N)\times D^s$$

$$U'(z_1,\cdots z_r,y) = (z_1\cdot j^m(U_y)(U_y^{-1}(sz_1,y)),\cdots$$
$$z_r\cdot j^m(U_y)(U_y^{-1}(sz_r,y)),\ y)$$

satisfies

$$U'(p_m^\infty(S(Z,P,N))\times D^s) = p_m^\infty(S(Z,P,N))\times D^s \quad .$$

PROOF of Lemma 53: This is an easy exercise using Lemma 42, which we leave to the reader.

LEMMA 54. Let Z, P, and N be as in Lemma 53. If

$$N\times D^s \xrightarrow{\omega} N\times D^s$$

is a diffeomorphism fibered over D^s, then the diffeomorphism

$$_rJ^m(\mathbb{R}\times P,N)\times D^s \xrightarrow{\omega'} {}_rJ^m(\mathbb{R}\times P,N)\times D^s$$

$$\omega'(z_1,\cdots w_r,y) = (j^m(\omega_y)(tz_1)\cdot z_1,\cdots$$

$$j^m(\omega_y)(tz_r)\cdot z_r,y)$$

satisfies

$$\omega'(p_m^\infty S(Z,P,N)\times D^s) = p_m^\infty S(Z,P,N)\times D^s$$

PROOF of Lemma 54: Use Lemma 44. QED

The sets S(Z,P,N) behave well under sunny collapsing:

LEMMA 55. Let

$$I\times P\times D^s \xrightarrow{F=(h,f,p_3)} I\times N\times D^s$$

be a fibered concordance, and let $F^u = (h^u, f^u, p_3)$ be an isotopy arising from a sunny collapse ϕ^u (as in Lemma 29 §I.C.) If $Z \subset {}_rJ^\infty$ is an IASCM of level m, then

$$(117)\quad {}_rj^\infty(f^u)^{-1}(S(Z,P,N)) = \Big\{ (t_1, x_1, \dots t_r, x_r, y) \in (I\times P)^{(r)}\times D^s \;\Big|\; (t_1\phi^u(x_1,y), x_1, \dots t_r\phi^u(x_r,y), x_r, y) \in {}_rj^\infty(f)^{-1}(S(Z,P,N)) \Big\}$$

and

$$(118)\quad {}_rj^m(f) \pitchfork p_m^\infty(S^*(Z,P,N) \implies {}_rj^m(f^u) \pitchfork p_m^\infty(S^*(Z,P,N))$$

for all u.

PROOF of Lemma 55. Fix u. Let $\mathcal{U}$ be the map

$$\mathbb{R}\times P\times D^s \longrightarrow \mathbb{R}\times P\times D^s$$

$$\mathcal{U}(t,x,y) = (t\phi^u(x,y), x, y) \quad .$$

By (39) $f^u = f\cdot\mathcal{U}$. Therefore, in the notation of Lemma 53

$$(119)\quad ({}_rj^m(f^u), p_2) = \mathcal{U}'\cdot({}_rj^m(f), p_2) \quad .$$

Now (117) is immediate from Lemma 53. Of course, (117) also holds with $A(Z)$ in place of Z, and therefore:

(120) (117) holds with $S^*(Z,P,N)$ in place of $S(Z,P,N)$.

For (118), note that

$$ {}_rj^m(f) \pitchfork p_m^\infty S^*(Z,P,N) \Rightarrow $$

$$ ({}_rj^m(f),p_2) \pitchfork p_m^\infty(S^*(Z,P,N))\times D^s \Rightarrow $$

$$ \mathcal{U}'\cdot({}_rj^m(f),p_2) \pitchfork \mathcal{U}'\cdot(p_m^\infty(S^*(Z,P,N))\times D^s) $$

(where $\mathcal{U}'$ is defined as in Lemma 53)

$$ \Rightarrow ({}_rj^m(f^u),p_2) \pitchfork p_m^\infty(S^*(Z,P,N))\times D^s \quad \text{by (119) and Lemma 53} $$

$$ \Rightarrow {}_rj^m(f^u) \pitchfork p_m^\infty(S^*(Z,P,N)) \ . \qquad \text{QED} $$

§II.B. Operations on Sets of Multijets.

The proof of Theorem D in Chapter III will require that whenever we use any invariant algebraic set Z and its associated $S(Z,P,N)$ we must also consider certain other sets Z' with their $S(Z',P,N)$'s. In this section we introduce four operations A, B, C, and D which make new algebraic sets Z' out of old algebraic sets Z, and we show that they take IASCMs to IASCMs. The definitions of the operations will mean more to someone who has read §D of the Introduction than to someone who has not.

NOTATION: If Z has rank r, write $I(Z)$ for the "index set" $\{1,\cdots r\}$.

§II.B.1. Operation A.

This has been defined already (Def. 49).

§II.B.2. Operation B.

This is a binary operation. Suppose Z and Z' are two algebraic sets. Suppose r is an integer and ϕ and ϕ' are injections

$$\phi : I(Z) \longrightarrow \{1,\cdots r\}$$

$$\phi': I(Z') \longrightarrow \{1,\dots r\}$$

such that $\phi(I(Z)) \cup \phi'(I(Z')) = \{1,\dots r\}$. Then given any $i \in I(Z)$ and $i' \in I(Z')$ define

$$B_{\phi,\phi',i,i'}(Z,Z') = \left\{ z \in {}_rJ^\infty \;\middle|\; \phi^* z \in Z, \right.$$

$$\left. \phi'^* z' \in Z',\ p_2 s z_{\phi(i)} = p_2 s z_{\phi'(i')} \right\}$$

Clearly this is an algebraic set of rank r and level $\max(\text{level }(Z), \text{level }(Z'))$. (Here ϕ^* is defined by $\phi^*(z_1,\dots) = (z_{\phi(1)},\dots)$.)

§II.B.3. Operation C (and C^0).

These are subtler than the preceding ones.

DEFINITION 56. Suppose $z \in p^\infty_{m+1}(Z)$, where $Z \subset {}_rJ^\infty$ is an algebraic set of level m. A tangent vector $v \in T_{sz}((\mathbb{C}^{p+1})^{(r)})$ is said to be along Z at z if, for some holomorphic map f

$$\mathbb{C}^{p+1} \supset \Omega \xrightarrow{\ f\ } \mathbb{C}^n$$

(Ω being open in $\mathbb{C}^{p+1}$ and containing sz_i for all i in $\{1,\ldots r\}$) such that

$$ {}_rj^{m+1}(f)(sz) = z \ , $$

we have

$$ [D({}_rj^m(f))(sz)]\cdot v \in T(p_m^\infty Z) \ . $$

REMARK 57: $T(p_m^\infty(Z))$ is the tangent space of the (possibly singular) variety $p_m^\infty(Z)$, in the sense of Zariski. (See [S] p.75).

REMARK 58: $D({}_rj^m(f))(sz)$ depends only on ${}_rj^{m+1}(f)(sz)$. Thus one could write "every" instead of "some" in Def. 56.

NOTATION: If $v \in T_{(x_1,\ldots x_r)}((\mathbb{C}^{p+1})^{(r)})$, write v_i for the "i-th component" of v in the obvious sense. Thus $v_i \in T_{x_i}(\mathbb{C}^{p+1})$, for $1 \le i \le r$.

Now suppose that Z is algebraic of level m and rank r, and that $i \in I(Z)$. Set

$$\tilde{C}_i^0(Z) = \left\{ (z,v,c) \;\middle|\; z \in p_{m+1}^{\infty}(Z)\ ,\ v \in T_{sz}((\mathbb{C}^{p+1})^{(r)}),\right.$$

$$\left. v \neq 0\ ,\ v \text{ is along } Z \text{ at } z,\ c \in \mathbb{C},\ v_i = c\frac{\partial}{\partial x_1}\right\}$$

$$\tilde{C}_i(Z) = \left\{ (z,v,c) \;\middle|\; z \in p_{m+1}^{\infty}(Z)\ ,\ v \in T_{sz}((\mathbb{C}^{p+1})^{(r)}),\right.$$

$$\left. v \neq 0\ ,\ v \text{ is along } Z \text{ at } z,\ c \in \mathbb{C},\ (Dz_i)\cdot(v_i - c\frac{\partial}{\partial x_1}) = 0\right\}$$

It is easy to see that $\tilde{C}_i^0(Z)$ and $\tilde{C}_i(Z)$ are Zariski closed sets in

$${}_rJ^{m+1} \times (T((\mathbb{C}^{p+1})^{(r)}) - (0\text{-section})) \times \mathbb{C} \ .$$

Now set

$$\left\{\begin{matrix} \hat{C}_i^0(Z) \\ \\ \hat{C}_i(Z) \end{matrix}\right\} = \text{image of } \left\{\begin{matrix} \tilde{C}_i^0(Z) \\ \\ \tilde{C}_i(Z) \end{matrix}\right\} \text{ in } {}_rJ^{m+1} \quad \text{under}$$

$$(z,v,c) \longmapsto z$$

Set

$$C_i^0(Z) = (p_{m+1}^{\infty})^{-1}\overline{(\hat{C}_i^0(Z))}$$

$$C_i(Z) = (p_{m+1}^{\infty})^{-1}\overline{(\hat{C}_i(Z))}$$

where the bar denotes Zariski closure or equivalently complex closure. (Fact 7, §I.A.2.). Clearly $C_i^0(Z)$ and $C_i(Z)$ are algebraic sets of rank r and level $m+1$.

§II.B.4. Operation D.

Suppose that Z is an algebraic set of rank r and level m and that ϕ is a surjection

$$I(Z) \xrightarrow{\phi} \{1,\ldots r'\}$$

but not a bijection. We will define an algebraic set $D_\phi(Z)$ of rank r' and level $\max_{i'=1}^{r'} ((m+1)|\phi^{-1}(i')| - 1)$.

To get a feeling for what D_ϕ does, the reader may want to look at Lemma 61.

First set $K = \mathrm{Poly}_{r\binom{p+1+m}{m}-1}(\mathbb{C}^{p+1})$. Thus K is a finite-dimensional complex vector space. The product K^n can be thought of as the space of all complex polynomial maps

$$\mathbb{C}^{p+1} \longrightarrow \mathbb{C}^n$$

of degree less than $r\binom{p+1+m}{m}$. Now set

$$\begin{aligned} X &= \left\{(x,f) \in (\mathbb{C}^{p+1})^{(r)} \times K^n \;\middle|\; {}_r j^{\infty}(f)(x) \in Z\right\} \\ &= \left\{(x,f) \in (\mathbb{C}^{p+1})^{(r)} \times K^n \;\middle|\; {}_r j^{m}(f)(x) \in p^{\infty}_{m}(Z)\right\} \end{aligned}$$

X is a Zariski-closed subset of $(\mathbb{C}^{p+1})^{(r)} \times K^n$. Set

$$Y = \left\{(y,f) \in (\mathbb{C}^{p+1})^{(r')} \times K^n \;\middle|\; (\phi^* y,f) \in \overline{X}\right\}.$$

($\phi^*(y_1,\dots y_{r'}) \underset{\mathrm{def}}{=} (y_{\phi(1)},\dots y_{\phi(r)})$. $\overline{X}$ is the Zariski closure, or equivalently the complex closure, of X in $(\mathbb{C}^{p+1})^r \times K^n$ (Fact 7, §I.A.2.).) Y is a Zariski-closed subset of $(\mathbb{C}^{p+1})^{(r')} \times K^n$. Set

$$D_{\phi}(Z) = \left\{(z_1,\dots z_{r'}) \in {}_{r'}J^{\infty} \;\middle|\; \text{for some } (y,f) \in Y, \text{ for all } i' \ (1 \le i' \le r'),\right.$$
$$\left. p^{\infty}_{(m+1)|\phi^{-1}(i')|-1}(z_{i'}) = j^{(m+1)|\phi^{-1}(i')|-1}(f)(y_{i'})\right\}.$$

The fact that $D_\phi(Z)$ is algebraic and of the indicated level will follow from

CLAIM 59. Y is a union of fibers for the map

$$(\mathbb{C}^{p+1})^{(r')} \times K^n \xrightarrow{j'} \prod_{i'=1}^{r'} J_{\mathbb{C}}^{(m+1)|\phi^{-1}(i')|-1}(\mathbb{C}^{p+1},\mathbb{C}^n) .$$

$$(y,f) \longrightarrow (j^{(m+1)|\phi^{-1}(1)|-1}(f)(y_1),\ldots$$

$$j^{(m+1)|\phi^{-1}(r')|-1}(f)(y_{r'}))$$

PROOF of Claim. For each i' $(1\le i'\le r')$ and each $x\in(\mathbb{C}^{p+1})^{(r')}$, define

$$W_{i'}(x) = \left\{ f\in K \;\middle|\; f \text{ vanishes to order } m+1 \text{ at } x_i \text{ for all } i\in\phi^{-1}(i')\right\} .$$

Set

$$W(x) = \bigcap_{i'=1}^{r'} W_{i'}(x) .$$

Lemma 36 (§I.E.1) implies that $W_{i'}(x)$ always has codimension $\binom{p+1+m}{m}|\phi^{-1}(i')|$ in K, and that W(x) has codimension $\binom{p+1+m}{m}$ r in K. Clearly, then, $W_{i'}$ and W are morphisms of algebraic varieties

$$(\mathbb{C}^{p+1})^{(r)} \longrightarrow \mathbb{G}_{i'}$$

$$(\mathbb{C}^{p+1})^{(r)} \longrightarrow \mathbb{G} ,$$

where $\mathbb{G}_{i'}$ and $\mathbb{G}$ are certain complex Grassmanian manifolds. Set

$$\tilde{X} = \left\{ (x,f,W,W_1,\dots W_{r'}) \in (\mathbb{C}^{p+1})^{(r)} \times K^n \times \mathbb{G} \times \mathbb{G}_1 \times \dots \times \mathbb{G}_{r'} \;\middle|\; (x,f) \in X, W = W(x), W_{i'} = W_{i'}(x) \quad \forall_{i'} \right\}$$

$$\tilde{Y} = \left\{ (y,f,W,W_1,\dots W_{r'}) \in (\mathbb{C}^{p+1})^{(r')} \times K^n \times \mathbb{G} \times \mathbb{G}_1 \times \dots \times \mathbb{G}_{r'} \;\middle|\; (\phi^* y,f,W,W_1,\dots W_{r'}) \in \bar{\tilde{X}} \right\} ,$$

where $\bar{\tilde{X}}$ is the closure (in either sense) of $\tilde{X}$ in

$$(\mathbb{C}^{p+1})^r \times K^n \times \mathbb{G} \times \mathbb{G}_1 \times \dots \times \mathbb{G}_{r'} .$$

Obviously $\tilde{X}$ projects isomorphically to X. Because the Grassmanians are projective (compact), it follows that $\bar{\tilde{X}}$ projects onto $\bar{X}$, and hence that $\tilde{Y}$ projects onto Y.

Clearly $W \subset W_{i'}$ for any $(x,f,W,W_1,\ldots W_{r'}) \in \tilde{X}$. Therefore the same is true if $(x,f,W,W_1,\ldots W_{r'}) \in \bar{\tilde{X}}$. Thus

$$(121) \qquad (y,f,W,W_1,\cdots W_{r'}) \in \tilde{Y} \Rightarrow W \subset W_{i'} \quad \forall_{i'} \, .$$

CLAIM 60: Let $(y,f,W,W_1,\cdots W_{r'}) \in \tilde{Y}$ and $1 \le i' \le r'$. Then any $h \in K$ which vanishes to order $(m+1)|\phi^{-1}(i')|$ at $y_{i'}$ is in $W_{i'}$.

PROOF of Claim 60: The polynomial h can be written

$$h(X) = \sum_{\alpha} a_\alpha \prod_{\phi(i)=i'} M_{\alpha,i}(X-y_{i'})$$

where $X = (X_1,\cdots X_{p+1})$ is a multivariable, α runs through some finite index set, $a_\alpha \in \mathbb{C}$, and the $M_{\alpha,i}$ are monomials satisfying

$$\deg M_{\alpha,i} \ge m+1$$

$$\sum_{\phi(i)=i'} \deg M_{\alpha,i} \le r\binom{p+1+m}{m} - 1$$

Define $h_x \in K$ for each $x \in (\mathbb{C}^{p+1})$ by

$$h_x(X) = \sum_\alpha a_\alpha \prod_{\phi(i)=i'} M_{\alpha,i}(X-x_i) .$$

Clearly $h_x \in W_{i'}(x)$ for any $x \in (\mathbb{C}^{p+1})^{(r)}$. Therefore, since $(\phi^* y, f, W, W_1, \dots W_{r'}) \in \bar{\tilde{X}}$ and h_x depends continuously on x, it follows that $h_{\phi^* y} \in W_{i'}$. But $h_{\phi^* y} = h$. QED

Now set

$$B = (\mathbb{C}^{p+1})^r \times \mathbb{G} \times \mathbb{G}_1 \times \dots \times \mathbb{G}_{r'}$$

$$E = (\mathbb{C}^{p+1})^r \times K^n \times \mathbb{G} \times \mathbb{G}_1 \times \dots \times \mathbb{G}_{r'} \quad .$$

E can be considered as a trivial algebraic vector bundle over B. $\tilde{X}$ is a closed subset of $E\big|_{(\mathbb{C}^{p+1})^{(r)} \times \mathbb{G} \times \dots \times \mathbb{G}_{r'}}$, $\bar{\tilde{X}}$ is a closed subset of E. The set

$$F \underset{\text{def}}{=} \left\{ (x, f, W, \dots W_{r'}) \in E \;\middle|\; f \in W^n \right\}$$

is a subbundle of E. Let $q\colon E \to E/F$ be the quotient map.

Then $\tilde{X} = q^{-1}q\tilde{X}$, since Z has level m. It follows that

(122) $$\bar{\tilde{X}} = q^{-1}q\bar{\tilde{X}} .$$

Now, to prove Claim 59 assume that $(y,f) \in Y$ and $j'(y,f) = j'(y,g)$. We must show that $(y,g) \in Y$. Choose $W, W_1,\ldots W_{r'}$ such that $(y,g,W,\ldots W_{r'}) \in \tilde{Y}$. By Lemma 36 the vector subspaces of K

$$\left\{h \in V \mid h \text{ vanishes to order } (m+1)\left|\phi^{-1}(i')\right| \text{ at } y_{i'}\right\}, \quad 1\le i'\le r$$

are independent, in the sense that the codimension of the intersection is the sum of the codimensions. Therefore by Claim 60 the subspaces $W_{i'}$ $(1\le i'\le r')$ are independent in the same sense. But $W \subset \bigcap_{i'=1}^{r'} W_{i'}$, by (121), and

$$\operatorname{codim}(W) = \sum_{i'=1}^{r'} \operatorname{codim}(W_{i'}) = \operatorname{codim}\left(\bigcap_{i'=1}^{r'} W_{i'}\right) .$$

It follows that

(123) $$W = \bigcap_{i'=1}^{r'} W_{i'} .$$

Now, $g-f$ vanishes to order $(m+1)|\phi^{-1}(i')|$ at $y_{i'}$ for all i', since $j'(y,f) = j'(y,g)$. Therefore each component of $g-f$ (an element of K) does the same. By Claim 60 these components are in $\bigcap_{i'=1}^{r'} W_{i'}$. So by (123) we have

$$g-f \in W^n . \tag{124}$$

Therefore,

$$q(\phi^* y,g,W,\ldots W_{r'}) = q(\phi^* y,f,W,\ldots W_{r'}) \in q\bar{\tilde{X}}$$

$$(\phi^* y,g,W,\ldots W_{r'}) \in \bar{\tilde{X}} \quad \text{by } (122)$$

$$(y,g,W,\cdots W_{r'}) \in \tilde{Y}$$

$$(y,g) \in Y.$$

This proves Claim 59. QED

LEMMA 61. Let Z be an algebraic set of level m and rank r. Let $\phi\colon I(Z) \longrightarrow \{1,\cdots r'\}$ be a surjection, and not a bijection. Let $\mathcal{U} \subset \mathbb{R}^{p+1}$ be open. Let

$\left\{f^{\nu}\right\}_{\nu=1}^{\infty}$ be a convergent sequence in $C^{\infty}(U,\mathbb{R}^n)$ with limit f. Let $\{x^{\nu}\}$ be a sequence in $U^{(r)}$ which converges to $x=\phi^{*}y$ in U^{r}, where $y \in U^{(r')}$. Then

$$_{r}j^{\infty}(f^{\nu})(x^{\nu}) \in Z \quad \forall_{\nu} \quad \Rightarrow \quad _{r'}j^{\infty}(f)(y) \in D_{\phi}(Z) .$$

PROOF of Lemma 61. Use Lemma 38. Lemma 38 clearly remains true if in its statement the C^{∞} functions f^{ν} and polynomials g^{ν} are allowed to take values in $\mathbb{R}^n$ instead of in $\mathbb{R}$. (Just apply the lemma separately for each coordinate in $\mathbb{R}^n$.) Now apply this generalized form of Lemma 38 to the case at hand, taking $p+1$ as the "m" of the statement and $m+1$ as the "k_i" for each i. This provides a sequence $\{g^{\nu}\}$ in K^n with a limit $g \in V^n$ such that:

(125) $\quad f^{\nu}-g^{\nu}$ vanishes to order $m+1$ at x_i^{ν} $\quad \forall_{\nu,i}$

(126) $f-g$ vanishes to order $(m+1)\left|\phi^{-1}(i')\right|$ at $y_{i'}$ for all i.

Statement (125) implies

(127) $\quad {}_{r}j^{m}(g^{\nu})(x^{\nu}) = {}_{r}j^{m}(f^{\nu})(x^{\nu}) \in p_m^{\infty}Z$ for all ν .

The sequence $\{(x^\nu, g^\nu)\}$ is thus in X (the "X" of the definition of operation D), so that

$$(x,g) \in \overline{X} .$$

It follows that

$$(y,g) \in Y$$

$$j'(y,g) \in j'(Y) .$$

But (126) implies that

$$j^{(m+1)|\phi^{-1}(i')|-1}(g)(y_{i'}) = j^{(m+1)|\phi^{-1}(i')|-1}(f)(y_{i'}),$$

so that $j'(y,f) = j'(y,g) \in j'(Y)$. Then by Claim 59 $(y,f) \in Y$, and ${}_{r'}j^\infty(f)(y) \in D_\phi(Z)$. QED

EXAMPLE 62: Let us see how the definition of $D_\phi(Z)$ works out in one very simple case: when $Z = Z^0$ (see Example 40, §II.A.1) and ϕ is the unique map $\{1,2\} \to \{1\}$. Thus, in the notation that was used in defining $D_\phi(Z)$, we have $r=2$, $m=1$, $r'=1$, $K = \mathrm{Poly}_1(\mathbb{C}^{p+1})$, $K^n = \mathrm{Poly}_1(\mathbb{C}^{p+1}, \mathbb{C}^n)$. Using $X_1, \cdots X_{p+1}$ as standard coordinates in $\mathbb{C}^{p+1}$, a polynomial $f \in K$ (resp. $f \in K^n$) has the form

$$f = a + \sum_{j=1}^{p+1} b_j X_j \ , \ a, b_j \in \mathbb{C}$$

$$(\text{resp.} \quad f = A + \sum_{j=1}^{p+1} B_j X_j \ , \ A, B_j \in \mathbb{C}^n) \ .$$

Now

$$X = \left\{ (x_1, x_2, A + \sum_j B_j X_j) \in (\mathbb{C}^{p+1})^{(2)} \times K^n \ \middle| \ A + \sum_j B_j x_{1,j} = A + \sum_j B_j x_{2,j} \right\}$$

An element $(y, A + \sum_j B_j X_j) \in \mathbb{C}^{p+1} \times K^n$ is in Y if and only if

$$(128) \qquad (y, y, A + \sum_j B_j X_j) \in \overline{X} \ .$$

CLAIM 63: (128) holds if and only if for some $v \neq 0$ in $\mathbb{C}^{p+1}$

$$\sum_j B_j v_j = 0 \tag{129}$$

PROOF of Claim: First suppose (129) for some $v \neq 0$. Then for every $\epsilon \neq 0$ in $\mathbb{C}$

$$(y, y + \epsilon v, A + \sum_j B_j X_j) \in X \quad .$$

Therefore (128) holds.

Conversely assume (128). Then some sequence $(x^\nu, x^\nu + v^\nu, A^\nu + \sum_j B_j^\nu X_j)$ in X has limit $(y, y, A + \sum_j B_j X_j)$. That is,

$$y = \lim_\nu x^\nu \quad \text{in} \quad \mathbb{C}^{p+1}$$

$$0 = \lim_\nu v^\nu \quad \text{in} \quad \mathbb{C}^{p+1}, \; v^\nu \neq 0 \quad \forall_\nu$$

$$A = \lim_\nu A^\nu \quad \text{in} \quad \mathbb{C}^n$$

$$B_j = \lim_\nu B_j^\nu \quad \text{in} \quad \mathbb{C}^n$$

$$\sum_j B_j^\nu v_j^\nu = 0 \qquad \forall_\nu$$

Write

$$v^\nu = |v^\nu| \cdot \hat{v}^\nu \ , \quad |\hat{v}^\nu| = 1 \quad .$$

After passing to a subsequence there is a limit

$$\lim_\nu \hat{v}^\nu = \hat{v} \neq 0 \quad .$$

Now

$$\sum_j B_j \hat{v}_j = \lim_\nu \sum_j B_j^\nu \hat{v}_j^\nu = \lim_\nu \frac{\sum_j B_j^\nu v_j^\nu}{|v_j^\nu|} = \lim_\nu 0 = 0 \quad .$$

Thus (129) holds, taking $v = \hat{v}$. QED

Therefore

$$Y = \left\{ (y, A + \sum_j B_j X_j) \in \mathbb{C}^{p+1} \times K^n \,\middle|\, \exists_{v \neq 0 \text{ in } \mathbb{C}^{p+1}} \sum_j B_j v_j = 0 \right\} .$$

j' is the map

$$\mathbb{C}^{p+1} \times K^n \longrightarrow J^1$$

$$(y,\ A + \sum_j B_jX_j) \longmapsto j^1(A + {}_{j}B_jX_j)(y)\ .$$

Thus

$$j'(Y) = \left\{ z \in J^1 \quad \ker(Dz) \neq 0 \right\}$$

$$Z^1 \underset{\text{def}}{=} D_\phi(Z^\circ) = \left\{ z \in J^\infty \ \middle|\ \ker(Dz) \neq 0 \right\}.$$

In other words, just as Z° detects double points Z^1 detects critical points: for any manifolds P^p and N^n we have

$$S(Z^1,P,N) = \left\{ z \in J^\infty(\mathbb{R}\times P,N) \ \middle|\ \ker Dz \neq 0 \right\}.$$

It is a fairly well-known principle in differential topology that critical points of a map f between manifolds can be thought of as a limiting case of double points. Thus if one wants to use the double-point set of f to make some geometrical construction one must usually be prepared to deal with the critical point set as well.

Operation D will allow us to treat "singular sets" other than the double-point set in much the same way, provided they are "algebraic".

§II.B.5. The operations preserve invariance.

The next result just says that the operations A through D preserve invariance of algebraic sets. It is almost obvious, but we have written out most of a proof.

LEMMA 64. If Z and Z' are IASCM, then the sets $A(Z)$, $B_{\phi,\phi',i,i'}(Z,Z')$, $C_i(Z)$, and $D_\phi(Z)$ (whenever they are defined) are also IASCM's.

PROOF of 64: In fact A-D preserve both domain-invariance and range-invariance. We will prove that they preserve domain invariance; range-invariance (which is easier) will be left to the reader.

A: Suppose Z is an algebraic set of rank r and level m and is domain-invariant. Given (100), consider the complex diffeomorphism $\mathcal{U}^*$, as defined by (101):

$$(130) \qquad \underset{\text{LHS}}{{}_rJ^m_{\mathbb{C}}(\Omega_2,\mathbb{C}^n)} \xrightarrow{\ \mathcal{U}^*\ } \underset{\text{RHS}}{{}_rJ^m_{\mathbb{C}}(\Omega_1,\mathbb{C}^n)}$$

Because Z is domain-invariant,

$$\mathcal{U}^*(p_m^\infty(Z) \cap \text{LHS}) = p_m^\infty(Z) \cap \text{RHS} .$$

Therefore

$$\mathcal{U}^*(p_m^\infty A(Z) \cap \text{LHS}) =$$

$$\mathcal{U}^*(\text{singular set of } p_m^\infty(\mathbb{Z}) \cap \text{LHS}) =$$

$$\text{singular set of } p_m^\infty(Z) \cap \text{RHS} =$$

$$p_m^\infty A(Z) \cap \text{RHS} .$$

(We have used the last part of Remark 8, §I.A.2., here.) Thus $A(Z)$ is domain-invariant

B: Let Z and Z' be algebraic sets of level m, and let $\phi,\phi',i,$ and i' be as required by operation B. Assume Z and Z' are domain-invariant. Given $z \in \text{LHS}$, we have

$$\mathcal{U}^* z \in p_m^\infty B_{\phi,\phi',i,i'}(Z,Z') \iff$$

$$\mathcal{U}^*\phi^* z \in p_m^\infty Z \text{ and } \mathcal{U}^*\phi'^* z \in p_m^\infty Z' \text{ and } \mathcal{V} p_2^{-1} s z_{\phi(i)} = \mathcal{V} p_2^{-1} s z_{\phi'(i')} \iff$$

$\phi^* z \in p_m^\infty Z$ and $\phi'^* z \in p_m^\infty Z'$ and $p_2 s z_{\phi(i)} = p_2 s z_{\phi'(i')} \Longleftrightarrow$

$$z \in p_m^\infty B_{\phi,\phi',i,i'}(Z,Z') .$$

Thus $B_{\phi,\phi',i,i'}(Z,Z')$ is domain-invariant.

C: Let Z be a domain-invariant algebraic set of rank r and level $m-1$ and let $i \in I(Z)$ (so that $C_i(Z)$ will have level m), and let $\mathcal{U}$ and $\mathcal{V}$ be as in (100). We have to show

$$\text{(131)} \qquad \mathcal{U}^*(p_m^\infty C_i(Z) \cap \text{LHS}) = p_m^\infty C_i(Z) \cap \text{RHS}$$

(where "LHS" and "RHS" as usual denote the two sides of (130)). To prove (131), consider the intermediate sets $\tilde{C}_i(Z)$ and $\hat{C}_i(Z)$ which occur in the definition of $C_i(Z)$. $\tilde{C}_i(Z)$ is a subset of

$$X \underset{\text{def}}{=} {}_rJ^m_{\mathbb{C}}(\mathbb{C}^{p+1},\mathbb{C}^n) \times T((\mathbb{C}^{p+1})^{(r)}) \times \mathbb{C} .$$

Define a holomorphic map $\tilde{\mathcal{U}}^*$ from

$$X_2 \underset{\text{def}}{=} {}_rJ^m_{\mathbb{C}}(\Omega_2,\mathbb{C}^n) \times T(\Omega_2{}^{(r)}) \times \mathbb{C}$$

to

$$X_1 \underset{\text{def}}{=} {}_rJ^m_{\mathbb{C}}(\Omega_1, \mathbb{C}^n) \times T(\Omega_1{}^{(r)}) \times \mathbb{C}$$

(X_1 and X_2 are open subsets of X) by

$$\tilde{\mathcal{U}}^*(z,v,c) = (\mathcal{U}^*(z), D((\mathcal{U}^r)^{-1})(sz).v, c\,\frac{\partial(p_i\mathcal{U})}{\partial x^1}(\mathcal{U}^{-1}(sz_i))$$

(Here $\mathcal{U}^r$ is the map $\underbrace{\mathcal{U}\times\ldots\times\mathcal{U}}_{r\text{ times}}$ from Ω_1^r to Ω_2^r.)

$\tilde{\mathcal{U}}^*$ is a complex diffeomorphism with inverse $\widetilde{\mathcal{U}^{-1*}}$.

CLAIM 65. $\tilde{\mathcal{U}}^*(\tilde{C}_i(Z) \cap X_2) \subset \tilde{C}_i(Z) \cap X_1$.

PROOF of Claim: Suppose $(z,v,c) \in \tilde{C}_i(Z) \cap X_2$. Let $(z',v',c') = \tilde{\mathcal{U}}(z,v,c)$. Because $z \in p_m^\infty(Z)$ and Z is domain-invariant we have

$$z' = \mathcal{U}^*(z) \in p_m^\infty(Z) \ . \tag{132}$$

Also v is along Z at z. This implies that v' is along Z at z', as we will now show. Since v is along

Z at z there exist Ω, an open subset of $\mathbb{C}^{p+1}$ containing sz_i for all $i \in I(Z)$, and a holomorphic map

$$f: \Omega \longrightarrow \mathbb{C}^n$$

such that

$$(133) \qquad {}_rj^m(f)(sz) = z$$

$$(134) \qquad (D({}_rj^{m-1}(f))(sz)).v \in T(p^\infty_{m-1}Z) .$$

Set

$$\Omega' = U^{-1}(\Omega \cap \Omega_2)$$

and let f' be the composition

$$\Omega' \xrightarrow{U|\Omega'} \Omega \cap \Omega_2 \hookrightarrow \Omega \xrightarrow{F} \mathbb{C}^n .$$

Then

$$\begin{aligned} {}_rj^m(f')(sz') &= {}_rj^m(f)(sz) \cdot {}_rj^m(U)(sz') \\ &= z \cdot {}_rj^m(U)((U^r)^{-1}(sz)) \text{ by (133)} \end{aligned}$$

$$= \mathcal{U}^*(z) = z'$$

and

$$(D({}_r j^{m-1}(f'))(sz')).v' =$$

$$(D({}_r j^{m-1}(f'))(sz')) \quad (D((\mathcal{U}^r)^{-1})(sz)).v =$$

$$\left\{ D\left[({}_r j^{m-1}(f'))\circ((\mathcal{U}^r)^{-1}) \right] (sz) \right\} .v =$$

$$\left\{ D\left[\mathcal{U}^* \circ {}_r j^{m-1}(f) \right] (sz) \right\} . v =$$

$$D(\mathcal{U}^*)\cdot(D({}_r j^{m-1}(f))(sz))\cdot v \in$$

$$D(\mathcal{U}^*)\cdot T(p^\infty_{m-1}(Z) \cap \text{LHS}), \text{ by (134)}$$

$$\subset T(p^\infty_{m-1}(Z) \cap \text{RHS}) \text{ because } Z \text{ is domain-invariant.}$$

Therefore v' is along Z at z'.

Finally,

$$(Dz_i)\cdot(v_i - c\frac{\partial}{\partial x_1}) = 0$$

so that

$$(Dz_i')\cdot(v_i'-c'\frac{\partial}{\partial x^1}) =$$

$$D(z_i\cdot j^m(\mathcal{U})(\mathcal{U}^{-1}(sz_i))).$$

$$\left[D(\mathcal{U}^{-1})(sz_i)\cdot v_i - c\,\frac{\partial(p_1\cdot\mathcal{U})}{\partial x^1}(\mathcal{U}^{-1}(sz_i))\cdot\frac{\partial}{\partial x^1}\right] =$$

$$D(z_i)\cdot\left[v_i - c\,\frac{\partial}{\partial x^1}\right] = 0 .$$

(For the second step observe that

$$(D\mathcal{U})\cdot\frac{\partial}{\partial x^1} = \frac{\partial(p_1\circ\mathcal{U})}{\partial x^1}\cdot\frac{\partial}{\partial x^1}$$

since (100) commutes.) This shows that (z',v',c') is in $P_m^\infty(Z)$ and completes the proof of Claim 65. QED

The Claim implies immediately that

$$\tilde{\mathcal{U}}^*(\tilde{C}_i(Z) \cap X_2) = \tilde{C}_i(Z) \cap X_1$$

(the reverse inclusion follows by applying the Claim to $\mathcal{U}^{-1}$).

It follows that

$$u^*(\hat{C}_i(Z) \cap \text{LHS}) = u^*(\text{image of } (\tilde{C}_i(Z) \cap X_2))$$

$$= \text{image of } \tilde{u}^*(\tilde{C}_i(Z) \cap X_2)$$

$$= \text{image of } (\tilde{C}_i(Z) \cap X_1)$$

$$= \hat{C}_i(Z) \cap \text{RHS}$$

so that

$$u^*(p_m^\infty C_i(Z) \cap \text{LHS}) = u^*\overline{(\hat{C}_i(Z) \cap \text{LHS})}$$

$$= \overline{\hat{C}_i(Z) \cap \text{RHS}}$$

$$= p_m^\infty C_i(Z) \cap \text{RHS} ,$$

where the bar denotes closure in LHS or RHS (in the usual complex topology). Thus $C_i(Z)$ is domain invariant.

D: Let Z be a domain-invariant algebraic set of rank r and level m, and let

$$\{1, \cdots r\} \xrightarrow{\phi} \{1, \cdots r'\}$$

be a surjection but not a bijection. To show that $D_\phi(Z)$ is domain-invariant, let m' be the level of $D_\phi(Z)$ and let (U,V) be as in (100). Suppose $z \in P^\infty_{m'} D_\phi(Z)$. This means, setting $y = sz \in (\mathbb{C}^{p+1})$, that there exists $f \in K^n$ such that

$$(135) \qquad (\phi^* y, f) \in \overline{X} \ , \quad \text{and}$$

$$(136) \quad j^{(m+1)|\phi^{-1}(i')|-1}(f)(sz_{i'}) = P^{m'}_{(m+1)|\phi^{-1}(i')|-1} z_{i'}$$

$$\text{for all } i' \text{ such that } 1 \le i' \le r \ .$$

(We use the notation of the definition of $D_\phi(Z)$.) Statement (135) means that there exist sequences x_ν in $(\mathbb{C}^{p+1})^{(r)}$ and f_ν in K^n such that

$$(137) \qquad \lim_\nu x_\nu = \phi^* y \text{ in } (\mathbb{C}^{p+1})^r$$

$$(138) \qquad \lim_\nu f_\nu = f \quad \text{in } K^n$$

$$(139) \qquad {}_r j^m(f_\nu)(x_\nu) \in P^\infty_m(Z) \quad \text{for all } \nu \ .$$

It can be assumed by passing to a subsequence that $x_\nu \in U^{(r)}$ for all ν, since $\phi^* y \in U^r$. Because Z is

Therefore:

$$(\mathcal{U}^r(x_\nu), g_\nu) \in X \quad \text{for all} \quad \nu, \quad \text{by (144)}$$

$$(\phi^*(\mathcal{U}^{r'}(y)), g) \in \bar{X} \quad \text{by (137) and (143)}$$

$$(\mathcal{U}^{r'}(y), g) \in Y$$

$$j'(\mathcal{U}^{r'}(y), g) \in j'(Y)$$

$$\mathcal{U}^*(z) \in p^\infty_{m'} D_\phi(Z) \quad \text{by (144)}$$

To find g_ν and g use Lemma 37 (§I.E.1), taking

$$(p+1, r, \{m+1\}, \Omega_2, \{f_\nu \circ \mathcal{U}^{-1}\}, \{\mathcal{U}((x_\nu)_i)\})$$

in place of

$$(m, r, \{k_i\}, \mathcal{U}, \{f^\nu\}, \{x_i^\nu\}).$$

The conclusions (68) and (69) of Lemma 37 are exactly (141) and (142).

This completes the proof that Operation D preserves domain-invariance.

domain-invariant, (139) then implies

$$(140) \qquad \mathcal{U}^*({}_r j^m(f_\nu)(x_\nu)) \in P_m^\infty(Z) \quad \text{for all} \quad \nu \quad .$$

The next step will be to find g_ν and g in K^n such that

$$(141) \quad {}_r j^m(g_\nu)(\mathcal{U}^r(x_\nu)) = \mathcal{U}^*({}_r j^m(f_\nu)(x_\nu)) \quad \text{for all} \quad \nu$$

$$(142) \quad j^{(m+1)|\phi^{-1}(i')|-1}(g)(\mathcal{U}(y_{i'})) = p^{m'}_{(m+1)|\phi^{-1}(i')|-1}\left[(\mathcal{U}^*({}_r j^{m'}(f)(y)))_{i'}\right]$$

for all i'

$$(143) \qquad \lim_\nu g_\nu = g \quad (\text{in} \quad K^n) \; .$$

It will follow that

$$(144) \qquad {}_r j^m(g_\nu)(\mathcal{U}^r(x_\nu)) \in P_m^\infty(Z)$$

by (140) and (141). Also by (136) and (142) we have

$$(145) \quad j^{(m+1)|\phi^{-1}(i')|-1}(g)(\mathcal{U}(y_{i'})) = p^{m'}_{(m+1)|\phi^{-1}(i')|-1}(\mathcal{U}^*(z)) \; .$$

§II.C. $\underline{Z}$

The purpose of this section is to introduce the particular IASCM's which will be used in the proof of Theorem D, to develop some notation for dealing with these sets, and to prove some of their properties.

DEFINITION 66: Z is the smallest collection of multijet sets which contains $Z°$ (see Example 40) and is closed under all of the operations A, B, C, D

§II.C.1 Formal Facts.

LEMMA 67. For every $Z \in Z$ there exist two equivalence relations $\sim$ and $\simeq$ on the set $I(Z)$, and a subset $\Delta \subset I(Z)$, such that

(146) $$i \sim j \implies p_2 s z_i = p_2 s z_j \quad \text{for all} \quad z \in Z$$

(147) $$i \simeq j \implies t z_i = t z_j \quad \text{for all} \quad z \in Z$$

(148) $$i \in \Delta \implies \ker(D z_i) \neq 0 \quad \text{for all} \quad z \in Z$$

(149) The smallest equivalence relation on $I(Z)$ which contains both $\sim$ and $\simeq$ is the total relation (i.e., the relation with one equivalence class).

(150) For each $i \in I(Z)$ either $i \in \Delta$ or there exists $j \in I(Z)$ such that $j \not\sim i \simeq j$.

PROOF of Lemma 67: If $Z = Z^\circ$, so that $I(Z) = \{1,2\}$, then define $\sim$, $\simeq$, and Δ by

$$1 \not\sim 2 \ , \ 1 \simeq 2 \ , \ \Delta = \emptyset \ .$$

These satisfy (146)-(150).

It now suffices to show that A-D preserve the property expressed in Lemma 67. This is obvious in the cases of A and C, because if Z' is either $A(Z)$ or $C_i(Z)$ then $Z' \subset Z$ so that the same $\sim$, $\simeq$, and Δ which work for Z will work for Z'.

For operation B, let $Z'' = B_{\phi,\phi',i,i'}(Z,Z')$ and suppose that $\left\{\begin{matrix} \sim, \simeq, \Delta \\ \sim', \simeq', \Delta' \end{matrix}\right\}$ satisfy (146)-(150) for $\left\{\begin{matrix} Z \\ Z' \end{matrix}\right\}$.
Define $\sim''$, $\simeq''$, and Δ'' as follows:

$\sim''$ is the smallest equivalence relation on $I(Z'')$ such that

$$\begin{cases} j_1 \sim j_2 \Longrightarrow \phi(j_1) \sim'' \phi(j_2) \\ j_1' \sim' j_2' \Longrightarrow \phi'(j_1') \sim'' \phi'(j_2') \\ \phi(i) \sim'' \phi'(i') \end{cases}$$

$\simeq''$ is the smallest equivalence relation on $I(Z'')$ such that

$$j_1 \simeq j_2 \Rightarrow \phi(j_1) \simeq'' \phi(j_2)$$

$$j_1' \simeq' j_2' \Rightarrow \phi'(j_1) \simeq'' \phi'(j_2)$$

$$\Delta'' = \phi(\Delta) \cup \phi'(\Delta') \quad .$$

Then $\sim''$, $\simeq''$, and Δ'' satisfy (146)-(150) for Z''.

Now let $Z' = D_\phi(Z)$, and assume that $\sim$, $\simeq$, and Δ satisfy (146)-(150) for Z. Define $\sim'$, $\simeq'$, and Δ' by

$\sim'$ is the smallest equivalence relation on $I(Z')$ such that $i \sim j \Longrightarrow \phi(i) \sim' \phi(j)$

$\simeq$ is the smallest equivalence relation on $I(Z')$ such that $i \simeq j \Longrightarrow \phi(i) \simeq' \phi(j)$

$$\Delta' = \phi(\Delta) \cup \left\{ i' \in I(Z') \;\middle|\; \begin{array}{l} \text{for some } i_1, i_2 \in \phi^{-1}(i') \\ i_1 \simeq i_2 \neq i_1 \end{array} \right\}$$

Then (146)-(150) can be checked for $\sim'$, $\simeq'$, and Δ' with respect to Z'. For (146), (147), or (148) suppose $z' \in Z$. That is (in the notation of the definition of $D_\phi(Z)$), there exists $f \in K^n$ such that

$$(151)\quad j^{(m+1)|\phi^{-1}(i')|-1}(f)(sz'_{i'}) = p^\infty_{(m+1)|\phi^{-1}(i')|-1}(z'_{i'}) \quad \text{for } 1 \le i' \le r'$$

and such that for some sequences x_ν in $(\mathbb{C}^{p+1})^{(r)}$ and f_ν in K^n we have

$$(152)\qquad \lim_\nu x_\nu = \phi^* sz'$$

$$(153)\qquad \lim_\nu f_\nu = f$$

$$(154)\qquad {}_r j^m(f_\nu)(x_\nu) \in p^\infty_m(Z) \quad \text{for all } \nu .$$

To check (146) for $\sim'$ it is enough to show that

$$i \sim j \implies p_2 sz'_{\phi(i)} = p_2 sz'_{\phi(j)} .$$

But if $i \sim j$ then

$$p_2(x_\nu)_i = p_2(x_\nu)_j \quad \text{for all } \nu$$

by (146) for $\sim$. Therefore by (152) in the limit

$$p_2((\phi^* sz')_i) = p_2((\phi^* sz')_j) ,$$

that is

$$p_2 sz'_{\phi(i)} = p_2 sz'_{\phi(j)} .$$

To check (147) for $\simeq'$ it is enough to show that

$$i \simeq j \quad \Rightarrow \quad tz'_{\phi(i)} = tz'_{\phi(j)}$$

But

$$i \simeq j \quad \Rightarrow \quad f_\nu((x_\nu)_i) = f_\nu((x_\nu)_j) \quad \text{for all } \nu \text{ by (147) for } \simeq$$

$$\Rightarrow \quad f(sz_{\phi(i)}) = f(sz_{\phi(j)}) \quad \text{by (152) and (153)}$$

$$\Rightarrow \quad tz_{\phi(i)} = tz_{\phi(j)} \quad \text{by (151)}$$

To check (148) for Δ', suppose $i' \in \Delta'$. Either there exists $i \in \Delta$ such that $i' = \phi(i)$, or there exist $i \neq j$ in $I(Z)$ such that $\phi(i) = \phi(j) = i'$ and $i \simeq j$. In the first case (148) for Δ implies

$$\ker(Df_\nu)((x_\nu)_i) \neq 0 \quad \text{for all} \quad \nu,$$

which by (152) and (153) implies

$$\ker(Df)(sz'_{i'}) \neq 0 ,$$

that is (by (151)),

$$\ker(Dz'_{i'}) \neq 0 .$$

In the second case (147) for $\simeq$ implies that

$$f_\nu((x_\nu)_i) = f_\nu((x_\nu)_j) \quad \text{for all} \quad \nu$$

$$(\text{and} \quad (x_\nu)_i \neq (x_\nu)_j) .$$

Since $(x_\nu)_i$ and $(x_\nu)_j$ both have the same limit $sz'_{i'}$ as $\nu \to \infty$, it follows that the limit function f cannot be an immersion at $sz'_{i'}$, that is

$$\ker(Df)(sz'_{i'}) \neq 0 .$$

The checking of (149) and (150) for $\sim'$, $\simeq'$, Δ', Z' is left to the reader. QED

DEFINITION 68. For each $Z \in \mathcal{Z}$ define equivalence relations $\sim_Z$ and $\simeq_Z$ on $I(Z)$ and a subset $\Delta_Z \subset I(Z)$ by

$$i \sim_Z j \iff \forall_{z \in Z}\ p_2 s z_i = p_2 s z_j$$

$$i \simeq_Z j \iff \forall_{z \in Z}\ t z_i = t z_j$$

$$i \in \Delta_Z \iff \forall_{z \in Z}\ \ker Dz_i \neq 0$$

LEMMA 69. For any $Z \in \mathcal{Z}$ $\sim_Z$, $\simeq_Z$, and Δ_Z satisfy the conditions (146)-(150) of Lemma 67.

PROOF of Lemma 69: Use Lemma 67 and note that any $\sim$, $\simeq$, and Δ which satisfy Lemma 67 are contained in $\sim_Z$, $\simeq_Z$, and Δ_Z respectively. This implies the conclusion.

QED

Next we introduce, as shorthand, a category whose set of objects is $\mathcal{Z}$.

DEFINITION 70. Let $Z, Z' \in \mathcal{Z}$. A <u>morphism</u> from Z to Z' is an injective map of sets

$$I(Z) \xrightarrow{\phi} I(Z')$$

such that $\phi^*(Z') \subset Z$. (Here if $\phi:\{1,\cdots r\} \to \{1,\cdots r'\}$ we write $\phi^*(z_1,\cdots z_{r'}) = (z_{\phi(1)},\cdots z_{\phi(r)})$.)

LEMMA 71. Definition 70 satisfies the usual axioms for a category. In this category every endomorphism is an automorphism.

PROOF of Lemma 71: The first statement is easy. For the second, observe that every endomorphism of Z is in particular a permutation of $I(Z)$, so that some positive power of it is the identity. QED

DEFINITION 72: If $Z \in \mathcal{Z}$ then denote the isomorphism class of Z by $[Z]$. For Z and Z' in $\mathcal{Z}$ write $[Z] \le [Z']$ if and only if there is a morphism from Z to Z'.

LEMMA 73: $\leq$ is a partial ordering of the set of isomorphism classes of $\mathcal{Z}$.

PROOF of Lemma 73: Use Lemma 71. QED

LEMMA 74: For any $Z \in \mathcal{Z}$ and $Z' \in \mathcal{Z}$ we have

$$[Z] \leq [A(Z)]$$

$$[Z] \leq [B_{\phi,\phi',i,i'}(Z,Z')]$$

$$[Z'] \leq [B_{\phi,\phi',i,i'}(Z,Z')]$$

$$[Z] \leq [C_i(Z)]$$

whenever the RHS is defined.

PROOF of Lemma 74: In each case a morphism must be exhibited. ϕ is a morphism from Z to $B_{\phi,\phi',i,i'}(Z,Z')$. ϕ' is a morphism from Z' to $B_{\phi,\phi',i,i'}(Z,Z')$. In each of the other cases the identity map of the set $I(Z)$ is a suitable morphism. QED

§II.C.2. Codimension.

Suppose that Z is an algebraic set of multijets, of level m and rank r. Then $p_m^\infty Z$ is a closed algebraic subset of the smooth variety ${}_rJ^m$. As such it has a codimension. Notice that $\operatorname{codim}(p_m^\infty Z, {}_rJ^m)$ is independent of m: for any $m'>m$ the map

$$ {}_rJ^{m'} \xrightarrow{p_m^{m'}} {}_rJ^m $$

is a submersion, so that

$$ \operatorname{codim}(p_m^\infty Z, {}_rJ^m) = \operatorname{codim}((p_m^{m'})^{-1} p_m^\infty Z, {}_rJ^{m'}) $$

$$ = \operatorname{codim}(p_{m'}^\infty Z, {}_rJ^{m'}) $$

DEFINITION 75: If Z is algebraic of level m and rank r then

$$ c(Z) = \operatorname{codim}(p_m^\infty Z, {}_rJ^m) - r(p+1) \quad . $$

(By the discussion above, $c(Z)$ is independent of m.)

$c(Z)$ is an important measure of the size of Z.

Eventually the sets $Z \in \mathcal{Z}$ will be used to define "singular sets" for parametrized families of concordances of P^p in N^n. For an s-parameter family in general position the singular set associated with Z will have dimension $s-c(Z)$. The next few lemmas have to do mainly with the growth of $c(Z)$; in particular we will show (Lemma 81) that for any integer $s \geq 0$ there are only finitely many Z such that $s-c(Z) \geq 0$. It is here that we use the assumption $n-p \geq 3$.

LEMMA 76: Let $[Z] \leq [Z']$. Then $c(Z) \leq c(Z')$. In fact either $c(Z) < c(Z')$ or rank (Z) = rank (Z'), and in the latter case Z' is isomorphic to a subset of Z.

PROOF of Lemma 76: Let ϕ be a morphism from Z to Z'. Define $\sim_{Z'}$, $\simeq_{Z'}$, and $\Delta_{Z'}$ according to Definition 68. For short call them $\sim$, $\simeq$, and Δ.

Observe that (149) and (150) of Lemma 69 imply that $I(Z')$ can be written as an increasing union

$$\phi(I(Z)) = I^{\circ} \subset I^1 \subset \ldots \subset I^h = I(Z')$$

for some $h \geq 0$, where for each k $(1 \leq k \leq h)$ either

$$(155)\qquad \begin{cases} I^k = I^{k-1} \cup \{i\} \\ i \notin I^{k-1}, \quad i \sim j \in I^{k-1}, \quad i \in \Delta, \quad \text{or} \end{cases}$$

$$(156)\qquad \begin{cases} I^k = I^{k-1} \cup \{i, i'\} \\ i, i' \notin I^{k-1}, \quad i \neq i' \simeq i \sim j \in I^{k-1}, \quad \text{or} \end{cases}$$

$$(157)\qquad \begin{cases} I^k = I^{k-1} \cup \{i\} \\ i \notin I^{k-1}, \quad i \simeq j \in I^{k-1} \quad . \end{cases}$$

To see this, set $I^o = \phi(I(Z))$. Suppose inductively that $I^o, \ldots I^a$ have been constructed in such a way that for each k $(1 \le k \le a)$ either (155), (156), or (157) holds. If $I^a = I(Z)$ then take $h=a$; there is nothing more to do. If $I^a \neq I(Z)$ then (149) implies that there is some $i \in I(Z)$, $i \notin I^a$, such that either

$$(158)\qquad i \simeq j \quad \text{for some} \quad j \in I^a, \text{ or}$$

$$(159)\qquad i \sim j \quad \text{for some} \quad j \in I^a .$$

Choose some such i. If it satisfies (158), then take

$I^{a+1} = I^a \cup \{i\}$ to get (157). If i satisfies (159) but not (158), then use (150) to get either (155) or (156).

For ease of notation renumber $I(Z')$ in such a way that for each k $(0 \le k \le h)$

$$I^k = \{1, \cdots r_k\} \quad \text{for some integer } r_k .$$

Thus $\operatorname{rank}(Z) = r_0 < r_1 < \cdots < r_h = \operatorname{rank}(Z')$.

Now inductively define for each k $(0 \le k \le h)$ an algebraic set $Z^{(k)}$ of multijets, of rank r_k , by:

$$Z^{(0)} = \left\{ (z_{\phi(1)}, \cdots z_{\phi(r_0)}) \;\middle|\; (z_1, \cdots z_{r_0}) \in Z \right\}$$

$$Z^{(k)} = \left\{ (z_1, \cdots z_{r_k}) \in {}_{r_k}J^{\infty} \;\middle|\; (z_1, \cdots z_{r_{k-1}}) \in Z^{(k-1)} \text{ and } \left(\begin{array}{c} \left\{ \begin{array}{l} \ker Dz_i \ne 0 \\ p_2 s z_i = p_2 s z_j \end{array} \right\} \text{ in case (155)} \\ \text{or} \\ \left\{ \begin{array}{l} tz_i = tz_{i'} \\ p_2 s z_i = p_2 s z_j \end{array} \right\} \text{ in case (156)} \\ \text{or} \\ tz_i = tz_j \quad \text{in case (157)} \end{array} \right) \text{ for } 1 \le k \le h \right\}$$

Then

(160) $$c(Z^{(k-1)}) < c(Z^{(k)}) \quad \text{for} \quad 1 \le k \le h \ .$$

In fact, in case (155) $c(Z^{(k)}) = c(Z^{(k-1)}) + n-p-1$, in case (156) $c(Z^{(k)}) = c(Z^{(k-1)}) + n-p-2$, and in case (157) $c(Z^{(k)}) = c(Z^{(k-1)}) + n-p-1$. Also $Z^{(h)} \supset Z'$, which implies

(161) $$c(Z^{(h)}) \le c(Z') \ .$$

Combining (160) and (161) we see that $c(Z) \le c(Z')$ and either $c(Z) < c(Z')$ or $h=0$. This completes the proof.

QED

LEMMA 77. For any $Z, Z' \in \mathcal{Z}$ (and suitable i, ϕ, etc.)

(i) $\operatorname{rank} A(Z) = \operatorname{rank} Z$

(ii) $\operatorname{level} A(Z) = \operatorname{level} Z$

(iii) $c(A(Z)) > c(Z)$

(iv) $c(B_{\phi,\phi',i,i'}(Z,Z')) \ge c(Z)$
with equality only if
$\operatorname{rank} B_{\phi,\phi',i,i'}(Z,Z') = \operatorname{rank} Z$

(v) $c(B_{\phi,\phi',i,i'}(Z,Z')) \geq c(Z')$

with equality only if

$\operatorname{rank} B_{\phi,\phi',i,i'}(Z,Z') = \operatorname{rank} Z'$

(vi) $\operatorname{level}(B_{\phi,\phi',i,i'}(Z,Z')) = \max(\operatorname{level} Z, \operatorname{level} Z')$

(vii) $\operatorname{rank} C_i(Z) = \operatorname{rank} Z$

(viii) $\operatorname{level} C_i(Z) = 1 + \operatorname{level} Z$

(ix) $c(C_i(Z)) > c(Z)$

(x) $c(D_\phi(Z)) > c(Z)$.

PROOF of Lemma 77: Only the statements involving c are new. Of these, (iii) is easy and (iv) and (v) follow from Lemmas 76 and 74. We save (ix) for last. For (x) use Claim 59 (§II.B.4): In the notation of the definition of $D_\phi(Z)$, we have

$$c(D_\phi(Z)) = \operatorname{codim}\left(j'(Y), \prod_{i'=1}^{r'} J_{\mathbb{C}}^{(m+1)|\phi^{-1}(i')|-1}(\mathbb{C}^{p+1}, \mathbb{C}^n)\right)$$

$$-r'(p+1)$$

$= \operatorname{codim}(Y, (\mathbb{C}^{p+1})^{r'} \times K^n) - r'(p+1)$

(by Claim 59, and because j' is a submersion)

$$= \dim(K^n) - \dim(Y)$$

$$\geq \dim(K^n) - \dim(\overline{X}-X)$$

(because Y is isomorphic to a subset of $\overline{X}$-X)

$$> \dim(K^n) - \dim(X)$$

$$= c(Z).$$

We now begin proving (ix). As a preliminary step we prove that $c(C_i^0(Z)) > c(Z)$. (This step is the only reason for introducing C_i^0 at all.) Let $Z \in \mathcal{Z}$ have rank r and level m, and choose $i \in I(Z)$. We have to show

(162) $$\dim p_{m+1}^{\infty} C_i^0(Z) < \dim p_{m+1}^{\infty} Z \ .$$

By definition $p_{m+1}^{\infty} C_i^0(Z)$ is the Zariski closure of $\hat{C}_i^0(Z)$ in ${}_rJ^{m+1}$, where $\hat{C}_i^0(Z)$ is the image of the algebraic variety $\tilde{C}_i^0(Z)$ under a certain morphism. It follows from Fact 6, §I.A.2, that $\hat{C}_i^0(Z)$ contains a Zariski dense open subset Ω^1 of $p_{m+1}^{\infty}(C_i^0(Z)$.

Suppose (162) fails. Then in particular $p^{\infty}_{m+1}C^0_i(Z)$ is not contained in the singular set $p^{\infty}_{m+1}A(Z)$, so Ω^1 intersects $p^{\infty}_{m+1}Z - p^{\infty}_{m+1}A(Z)$ in a nonempty open subset Ω^2. Thus Ω^2 is a nonempty open subset of the nonsingular variety $p^{\infty}_{m+1}Z - p^{\infty}_{m+1}A(Z)$, and also a subset of $\hat{C}^0_i(Z)$.

Let $d \geq 0$ be an integer (to be chosen very large eventually) and consider the vector space of polynomial maps $V_d \underset{\text{def}}{=} \text{Poly}\,(\mathbb{C}^{p+1},\mathbb{C}^n)$. Define a "universal (r,m+1)-multijet map" ${}_rJ^{m+1}_d$ for maps in V_d:

$$(\mathbb{C}^{p+1})^{(r)} \times V_d \xrightarrow{{}_rJ^{m+1}_d} {}_rJ^{m+1}$$

$$(x_1,\cdots x_r,f) \longmapsto {}_rj^{m+1}(f)(x_1,\cdots x_r)$$

CLAIM 78: If d is large enough then ${}_rJ^{m+1}_d$ is a submersion and a surjection.

PROOF of Claim: Use Lemma 36 (§I.E.1). QED.

Now, choosing d according to the Claim, in particular ${}_rJ^{m+1}_d$ is transverse to Ω^2. Set $\Omega^2_d = \left({}_rJ^{m+1}_d\right)^{-1}\left(\Omega^2\right)$. Thus Ω^2_d is a nonempty complex submanifold of $(\mathbb{C}^{p+1})^{(r)} \times V_d$ with codimension $r(p+1) + c(Z)$ (the codimension of Ω^2 in ${}_rJ^{m+1}$).

Consider the projection

$$\Omega_d^2 \xrightarrow{\pi_i} \mathbb{C}^p \times V_d$$

$$(x_1, \cdots x_r, f) \longmapsto (p_2 x_i, f)$$

CLAIM 79: π_i is nowhere an immersion.

PROOF of Claim: Let $(x_1, \cdots x_r, f) \in \Omega_d^2$. Then

$$_r j^{m+1}(f)(x_1, \cdots x_r) \in \Omega^2 \subset \hat{C}_i^0(Z),$$

so by definition of $\hat{C}_i^0(Z)$ there exist $a \in \mathbb{C}$ and $v \in T_{(x_1, \cdots x_r)}((\mathbb{C}^{p+1})^{(r)})$ such that

(163) $$v \neq 0$$

(164) v is along Z at $_r j^{m+1}(f)(x_1, \cdots x_r)$

(165) $$a \frac{\partial}{\partial x^1} = v_i$$

(164) means that

$$(D_r J_d^{m+1})\cdot(v,0) = D({}_r j^{m+1}(f))\cdot v$$

$$\in T_{{}_r j^{m+1}(f)(x_1,\cdots x_r)}(p_{m+1}^{\infty} Z - p_{m+1}^{\infty} A(Z))$$

$$= T_{{}_r j^{m+1}(f)(x_1,\cdots x_r)}(\Omega^2) ,$$

in other words that

$$(v,0) \in T_{(x_1,\cdots x_r,f)}(\Omega_d^2) .$$

By (163) and (165) $(v,0)$ is a nontrivial vector in the kernel of $D\pi_i$. Therefore π_i is not an immersion at $(x_1,\cdots x_r,f)$. This proves the Claim.

Claim 79 implies (Fact 11, §I.A.2) that all of the nonempty fibers of π_i have positive dimension. Therefore

(166) $$\dim \left\{(a_1,a_2) \in \Omega_d^2 \times \Omega_d^2 \;\middle|\; a_1 \neq a_2, \pi_i a_i = \pi_i a_2\right\} > \dim \Omega_d^2 .$$

This can be expressed in terms of the sets $B_{\phi,\phi',i,i'}(Z,Z)$.

Let (ϕ, ϕ', r'') be such that ϕ and ϕ' are injections

$$\{1,\ldots r\} \longrightarrow \{1,\ldots r''\}$$

with Image (ϕ) $\cup$ Image $(\phi') = \{1,\cdots r''\}$. Consider the algebraic set (of level m)

$$B \underset{\text{def}}{=} B_{\phi,\phi',i,i}(Z,Z) .$$

Map the set

$$({}_{r''}J_d^{m+1})^{-1}(p_{m+1}^{\infty}B) \subset (\mathbb{C}^{p+1})^{(r'')} \times V_d$$

to $((\mathbb{C}^{p+1})^{(r)} \times V_d) \times ((\mathbb{C}^{p+1})^{(r)} \times V_d)$ by

$$(x,f) \longmapsto ((\phi^* x, f), (\phi'^* x, f)).$$

The image of this map lies in the set

$$\left\{(a_1, a_2) \in ({}_{r''}J_d^{m+1})^{-1}(p_{m+1}^{\infty}Z) \times ({}_{r''}J_d^{m+1})^{-1}(p_{m+1}^{\infty}Z) \;\middle|\; a_1 \neq a_2, \pi_i a_1 = \pi_i a_2 \right\},$$

and in fact the union of the various images (corresponding to various choices of (ϕ, ϕ', r'')) is this whole set.

Therefore the union of the images contains the LHS of (166). Then (166) implies that for some (ϕ,ϕ',r'')

$$\text{(167)} \qquad \dim(({}_{r''}J_d^{m+1})^{-1}(p_{m+1}^{\infty}B)) > \dim \Omega_d^2 .$$

But by Claim 78 d can be chosen large enough so that

$$(\mathbb{C}^{p+1})^{(r'')} \times V_d \xrightarrow{{}_{r''}J_d^{m+1}} {}_{r''}J^{m+1}$$

is transverse to all of the strata

$$p_{m+1}^{\infty}A^{\nu}(B) - p_{m+1}^{\infty}A^{\nu+1}(B)$$

of $p_{m+1}^{\infty}(B)$, for all possible B (there are only finitely many B). Then

$$\begin{aligned}
\dim({}_{r''}J_d^{m+1})^{-1}(p_{m+1}^{\infty}B) &= \dim V_d + r''(p+1) - \operatorname{codim}(p_{m+1}^{\infty}B, {}_{r''}J^{m+1}) \\
&= \dim V_d - c(B) \\
&\le \dim V_d - c(Z) \\
&= \dim({}_{r}J_d^{m+1})^{-1}(p_{m+1}^{\infty}Z) \\
&= \dim \Omega_d^2 ,
\end{aligned}$$

a contradiction to (167). (The inequality in the third step comes from Lemma 77(iv).) This completes the proof that $c(C_i^0(Z)) > c(Z)$.

To prove that $c(C_i(Z)) > c(Z)$ we begin as before: Suppose the contrary, so that

$$\dim p_{m+1}^{\infty} C_i(Z) = \dim p_{m+1}^{\infty} Z.$$

Conclude that there is a nonempty open set

$$\Omega^3 \subset p_{m+1}^{\infty}(Z) - p_{m+1}^{\infty} A(Z)$$

contained in $\hat{C}_i(Z)$. Set

$$\Omega^4 = \Omega^3 \cap p_{m+1}^{\infty}(Z - C_i^0(Z)).$$

By what has already been proved Ω^4 is not empty. As in the earlier argument we will use some V_d. Choose d large enough (using Claim 78) to insure that ${}_rJ_d^{m+1}$ is a surjective submersion. Then

$$\Omega_d^4 \underset{\text{def}}{=} ({}_rJ_d^{m+1})^{-1}(\Omega^4)$$

is a nonempty complex submanifold of $(\mathbb{C}^{p+1})^{(r)} \times V_d$ of codimension $r(p+1) + c(Z)$.

CLAIM 80: The map

$$\Omega_d^4 \xrightarrow{\pi_i} \mathbb{C}^p \times V_d$$

$$(x_1, \cdots x_r, f) \longmapsto (p_2 x_i, f)$$

is an immersion.

PROOF of Claim: This is more or less the reverse of the proof of Claim 79. Suppose π_i fails to be an immersion at some point $(x_1, \cdots x_r, f) \in \Omega_d^4$. Then there is some nonzero vector

$$v \in T_{(x_1, \cdots x_r)}(\mathbb{C}^{p+1})^{(r)}$$

such that

(168) $$(v, 0) \in T_{(x_1, \cdots x_r, f)}(\Omega_d^4) ,$$

and such that for some $a \in \mathbb{C}$

$$v_i = a\frac{\partial}{\partial x^1} \quad .$$

Statement (168) means that

$$(D_r J_d^{m+1})\cdot(v,0) \in T_{{}_r j^{m+1}(f)(x_1,\cdots x_r)}(\Omega^4)$$

$$= T_{{}_r j^{m+1}(f)(x_1,\cdots x_r)}(p_{m+1}^{\infty} Z);$$

thus v is along Z at ${}_r j^{m+1}(f)(x_1,\cdots x_r)$, and

$$({}_r j^{m+1}(f)(x_1,\cdots x_r),v,a) \in \tilde{C}_i^0(Z)$$

$${}_r j^{m+1}(f)(x_1,\cdots x_r) \in \hat{C}_i^0(Z) \subset p_{m+1}^{\infty} C_i^0(Z)$$

$${}_r j^{m+1}(f)(x_1,\cdots x_r) \notin \Omega^4$$

$$(x_1,\cdots x_r,f) \notin \Omega_d^4 \ ,$$

a contradiction. QED.

Now consider the algebraic vector bundle $T(\Omega_d^4) \times \mathbb{C}$ over Ω_d^4. Define a subset S of this bundle by letting the intersection of S with the fiber above $(x_1,\cdots x_r,f) \in \Omega_d^4$ be

$$\left\{(v,w,a) \in T_{(x_1,\cdots x_r,f)}(\Omega_d^4) \times \mathbb{C} \subset T_{(x_1,\cdots x_r)}((\mathbb{C}^{p+1})^{(r)}) \times T_f V_d \times \mathbb{C} \;\middle|\; w = 0,\ (Df)(x_i)\cdot(v_i - a\frac{\partial}{\partial x^1}) = 0\right\}$$

The set S is a Zariski closed subspace and meets each fiber in a vector subspace. For each fiber this subspace has positive dimension because $\Omega^4 \subset \hat{C}_i(Z)$. Let $\Omega_d^5 \subset \Omega_d^4$ be the subset where this dimension is minimal. Ω_d^5 is Zariski open in Ω_d^4 by Fact 9, §I.A.2. It follows from Fact 10 that over Ω_d^5 S is a vector subbundle of positive rank. On some still smaller nonempty open set $\Omega_d^6 \subset \Omega_d^5$, then, S has a nowhere-vanishing section $(v,0,a)$. Then $(v,0)$ is an algebraic (in particular, holomorphic) vector field on Ω_d^6. We are going to obtain a contradiction by using the integral curves of $(v,0)$.

Suppose that $\mathcal{U} \subset \mathbb{C}$ is an open set containing zero, and that γ and f are holomorphic maps

$$\mathcal{U} \xrightarrow{\ \gamma\ } \mathbb{C}^{p+1} \xrightarrow{\ f\ } \mathbb{C}^n .$$

Consider the holomorphic map

$$\mathcal{U} \xrightarrow{\mathcal{D}_{\gamma,f}} \bigwedge^2(\mathbb{C}^n)$$

$$\mathcal{D}_{\gamma.f}(t) = \left[(Df)(\gamma(t))\cdot\gamma'(t)\right] \wedge \left[(Df)(\gamma(t))\cdot\frac{\partial}{\partial x^1}\right].$$

Let $k \geq 0$ be an integer (to be chosen large eventually). The k-jet $j^k(\mathcal{D}_{\gamma.f})(0)$ depends only on the $(k+1)$-jets

$$j^{k+1}(\gamma)(0) \in J^{k+1}_{\mathbb{C}}(\mathbb{C},\mathbb{C}^{p+1})$$

$$j^{k+1}(f)(\gamma(0)) \in J^{k+1}_{\mathbb{C}}(\mathbb{C}^{p+1},\mathbb{C}^n) ,$$

and it depends on them polynomially. Set

$$X_k = \left\{ (u,z) \in J^{k+1}_{\mathbb{C}}(\mathbb{C},\mathbb{C}^{p+1}) \times J^{k+1}_{\mathbb{C}}(\mathbb{C}^{p+1},\mathbb{C}^n) \;\middle|\; \begin{array}{l} \text{for some } \mathcal{U},\ \gamma, \text{ and } f \text{ as above} \\ u = j^{k+1}(\gamma)(0),\quad z = j^{k+1}(f)(\gamma(0)) , \\ j^k(\mathcal{D}_{\gamma,f})(0) = 0,\quad \gamma'(0) \wedge \frac{\partial}{\partial x^1} \neq 0 \end{array} \right\}.$$

X_k is an algebraic variety. (It is defined by Zariski open and closed conditions in $J^{k+1}_{\mathbb{C}}(\mathbb{C},\mathbb{C}^{p+1}) \times J^{k+1}_{\mathbb{C}}(\mathbb{C}^{p+1},\mathbb{C}^n)$.)

Now Let Γ be an integral curve for the vector field $(v,0)$ in Ω_d^5. Thus Γ is a holomorphic map

$$\mathcal{U} \xrightarrow{\ \Gamma\ } \Omega_d^6 \subset (\mathbb{C}^{p+1})^r \times V_d$$

defined in some neighborhood $\mathcal{U}$ of 0 in $\mathbb{C}$. Write

$$\Gamma(t) = (\gamma_1(t), \ldots \gamma_r(t), f) \ .$$

(The component in V_d is constant because the vector field has the form $(v,0)$.) The map

$$\mathcal{U} \xrightarrow{\ \mathcal{D}_{\gamma_i,f}\ } \Lambda^2(\mathbb{C}^n)$$

is identically zero, because for any $t \in \mathcal{U}$

$$\mathcal{D}_{\gamma_i,f}(t) = \left[(Df)(\gamma_i(t))\cdot v_i(\Gamma(t))\right] \wedge \left[(Df)(\gamma_i(t))\cdot\frac{\partial}{\partial x^1}\right]$$

$$= a(\Gamma(t)) \left[(Df)(\gamma_i(t))\cdot\frac{\partial}{\partial x^1}\right] \wedge \left[(Df)(\gamma_i(t))\cdot\frac{\partial}{\partial x^1}\right]$$

$$= 0 \ .$$

Therefore for any $k \geq 0$ the jet $j^k(\mathcal{D}_{\gamma_i,f})(0)$ is zero, and the pair $(j^{k+1}(\gamma_i)(0),\ j^{k+1}(f)(\gamma_i(0)))$ is in X_k. (The last condition in the definition of X_k requires that $v_i(\Gamma(0))$ is linearly independent of $\frac{\partial}{\partial x^1}$, which is true by Claim 80.) In particular $j^{k+1}(f)(\gamma_i(0))$ is in the image of X_k under projection to the second factor. But every point in Ω_d^6 lies on some integral curve. Thus:

(169) $j^{k+1}(f)(x_i)$ is in the image of X_k for any

$$(x_1, \cdots x_r, f) \in \Omega_d^6 .$$

It is not hard to compute the codimension of X_k:

$$\operatorname{codim}(X_k, J_{\mathbb{C}}^{k+1}(\mathbb{C}, \mathbb{C}^{p+1}) \times J_{\mathbb{C}}^{k+1}(\mathbb{C}^{p+1}, \mathbb{C}^n)) =$$

$$1 + (p+1) + (k+1)(n-1) .$$

Therefore the codimension in $J_{\mathbb{C}}^{k+1}(\mathbb{C}^{p+1}, \mathbb{C}^n)$ of the Zariski closure of the image of X_k is at least

$$1 + (p+1) + (k+1)(n-1) - \dim J_{\mathbb{C}}^{k+1}(\mathbb{C}, \mathbb{C}^{p+1}) = (k+1)(n-p-2) .$$

Whatever k is, d can be chosen large enough so that

$$\mathbb{C}^{p+1} \times V_d \xrightarrow{J_d^{k+1}} J^{k+1}$$

is a submersion (Claim 78). In particular, then, by (169) we have

$$\operatorname{codim}(\Omega_d^6, (\mathbb{C}^{p+1})^{(r)} \times V_d) \geq (k+1)(n-p-2) .$$

That is,

$$c(Z) + r(p+1) \geq (k+1)(n-p-2) .$$

But k was arbitrary, and $n-p-2 \geq 1$. QED

Now all the necessary ingredients are ready for proving the following crucial property of $\mathcal{Z}$:

LEMMA 81: For any integer N there are only finitely many sets $Z \in \mathcal{Z}$ such that $c(Z) \leq N$.

PROOF of Lemma 81: From Def. 66 and parts (iii), (iv), (v), (ix), (x) of Lemma 77 it is clear that

$$c(Z) \geq c(Z^0) \quad \text{for all} \quad Z \in \mathcal{Z} \; .$$

Thus if Lemma 81 fails there must be a smallest integer N for which it fails. Assume this is the case.

It is clear from Def. 66 that the members of $\mathcal{Z}$ can be listed in an infinite sequence

$$Z^0, Z^1, \ldots$$

such that

(170) For all $\nu > 0$ either

$$\left\{ \begin{array}{l} Z^\nu \text{ is obtained from some } Z^\mu, \; 0 \leq \mu < \nu \; , \\ \text{by an operation } A, C, \text{ or } D \end{array} \right.$$

or

$$\left\{ \begin{array}{l} Z^\nu \text{ is obtained from some } Z^{\mu_1} \text{ and } Z^{\mu_2}, \\ 0 \leq \mu_i < \nu, \text{ by an operation } B \; . \end{array} \right.$$

Delete from this sequence all Z^ν such that $c(Z^\nu) > N$, and again call the sequence Z^ν. Then by (iii), (iv), (v), (ix), (x) of Lemma 77 the new sequence still satisfies (170), and by assumption it is still an infinite sequence. By (ix) and (x) of Lemma 77, whenever $Z^\nu = C_i(Z^\mu)$ or $Z^\nu = D_\phi(Z^\mu)$

in (170) then $c(Z^{\mu}) < N$. By minimality of N this can occur for only finitely many μ, hence for only finitely many ν. From this it follows (using (ii) and (vi) of Lemma 77) that level (Z^{ν}) is bounded by some m independent of ν.

Now extract a subsequence $\left\{Z^{\nu_j}\right\}_{j=0}^{\infty}$ such that for each j $Z^{\nu_{j+1}}$ is obtained from Z^{ν_j} by an operation A, C, or D, or from Z^{ν_j} and some Z^{μ} by an operation B. By deleting a finite number of terms from the beginning of the subsequence we can assume that operation D does not occur, and also that $c(Z^{\nu_j}) = N$ for all j. From (i), (iv), (v), and (vii) of Lemma 77 it follows that rank (Z^{ν_j}) is independent of j. Thus Z^{ν_j} is a sequence in Z of constant rank r and level m, which by Lemma 74 admits morphisms

$$Z^{\nu_1} \xrightarrow{\phi_1} Z^{\nu_2} \xrightarrow{\phi_2} Z^{\nu_3} \longrightarrow \cdots$$

This means that the variety ${}_rJ^m$ has an infinite descending chain of Zariski closed subsets

$$\left\{ \phi_1^*\phi_2^*\cdots\phi_{j-1}^*(p_m^{\infty}Z^{\nu_j}) \right\}_{j=1}^{\infty} .$$

Noetherian-ness (Fact 5, §I.A.2) implies that the chain eventually terminates. That is, for all large j ϕ_j is an isomorphism. But the Z^{ν_j} are distinct, and obviously each isomorphism class of IASCM's is finite, so this is impossible. QED.

Now list the members of Z:

$$Z_0, Z_1, Z_2, \ldots$$

in such a way that

(171) $$[Z_\alpha] \lneq [Z_\beta] \Rightarrow \alpha < \beta$$

(172) $$\alpha \leq \beta \Rightarrow c(Z_\alpha) \leq c(Z_\beta).$$

(Lemmas 81 and 76 imply that it is possible to do so.) This way of listing the members of Z will be fixed from now on.

It will be convenient to take

$$Z_0 = Z^0$$

$$Z_1 = D_\phi(Z^0) \qquad (\phi:\{1,2\} \rightarrow \{1\}).$$

(It is not hard to see that this is consistent with (171) and (172).)

§II.C.3. Goodness and Badness.

The rest of §C is concerned with certain special members of Z ($\hat{Z}_k$ for $k\geq 1$, and $\hat{Z}$), which will be used to define which elements of Z are "good" and which are "bad". (Def. 84 below thus corresponds to the definition at the very end of the Introduction.)

DEFINITION 82: For each integer $k\geq 1$ define

$$\hat{Z}_k = \left\{ (z_1,\dots z_{2k}) \in {}_{2k}J^\infty \;\middle|\; p_2sz_1 = p_2sz_2, tz_2 = tz_3, \right.$$

$$\left. p_2sz_3 = p_2sz_4, \dots\; tz_{2k} = tz_1 \right\}.$$

Define

$$\hat{Z} = \left\{ z \in {}_1J^\infty \;\middle|\; (Dz)\cdot\frac{\partial}{\partial x^1} = 0 \right\}.$$

LEMMA 83: $\hat{Z}_k \in Z$, $\hat{Z} \in Z$, $c(\hat{Z}_k) = k(n-p-2)$, and $c(\hat{Z}) = n-p-1$.

PROOF of Lemma 83: $\hat{Z}_k$ can be obtained from Z_0 by repeated use of operation B. For example,

$$\hat{Z}_1 = B_{\phi,\phi,1,2}(Z_0,Z_0), \quad \text{where}$$

$$\phi\colon \{1,2\} \to \{1,2\} \quad \text{is the identity}$$

$\hat{Z}_2$ requires two steps; for example,

$$\hat{Z}_2 = B_{\phi',\phi',3,4}(Z',Z'), \quad \text{where}$$

$$\phi'\colon \{1,2,3,4\} \to \{1,2,3,4\} \quad \text{is the identity and}$$

$$Z' = B_{\phi'',\phi''',1,2}(Z_0,Z_0) ,$$

$$\phi''(1) = 2 \quad \phi''(2) = 3$$

$$\phi'''(1) = 4 \quad \phi'''(2) = 1 .$$

In general $\hat{Z}_k$ requires k steps. We leave the details to the reader. Also,

$$\hat{Z} = D_\phi(\hat{Z}_1), \quad \text{where} \quad \phi\colon \{1,2\} \to \{1\}, \quad \text{by an}$$

argument like the one in Example 62 (§II.B.4). Details are again left as an exercise, as are the computations of codimensions. QED.

DEFINITION 84: An element $Z \in \mathcal{Z}$ is <u>bad</u> if either $[\hat{Z}_k] \le [Z]$ for some k or $[\hat{Z}] \le [Z]$. Otherwise Z is <u>good</u>.

LEMMA 85: If $Z \in \mathcal{Z}$ is bad then $c(Z) \ge n-p-2$.

PROOF of Lemma 85: Immediate from Lemmas 83 and 76.

QED

§II.D. Singular Sets for Fibered Concordances.

Now that the collection $\mathcal{Z}$ has been introduced, the general position principles of §I.D can be applied to it. This section begins with a definition of "general position with respect to $\mathcal{Z}$" for a fibered concordance, and a proof that every fibered concordance F can be put into general position by a small fibered isotopy. In the remainder of the section some facts and notation are developed concerning the "singular sets" $S_\alpha(F)$ which then arise as transverse preimages.

In particular we define sets $W_\alpha^{(D,R)}(F)$ in terms of the $\{S_\alpha(F)\}$. The chapter ends with a series of lemmas which assert in particular that for any F in general position the images $p_{2,3}\pi_i W_\alpha^{(D,R)}(F)$ in $\text{int}\, P \times D^s$ are the (positive-codimensional) "strata" of something which is very much like a stratification of int $P \times D^s$. (The only respect in which they fail to form a stratification is the difficulty discussed in Remark 89 below.)

§II.D.1. <u>General Position with Respect to Z.</u>

If F is a fibered concordance of P in N over D^s

$$I \times P \times D^s \xrightarrow{F = (h,f,p_3)} I \times N \times D^s$$

then the map f

$$I \times P \times D^s \xrightarrow{f} N$$

can be viewed as a family of maps from $I \times P$ to N parametrized by D^s, as in §I.D. Thus for any $r \geq 1$ and $m \geq 0$ there is a map

$$(I \times P)^{(r)} \times D^s \xrightarrow{{}_r\tilde{j}^m(f)} {}_rJ^m(I \times P, N) \ .$$

DEFINITION 86. Let Z be an IASCM of rank r and level m. A fibered concordance $F = (h,f,p_3)$ of P in N over D^s is <u>in general position</u> with respect to Z if the composition

$$(173) \qquad (I \times \mathrm{int}P)^{(r)} \times D^s \longrightarrow (I\times P)^{(r)} \times D^s$$

$$\xrightarrow{\;{}_r\tilde{j}^m(f)\;} {}_rJ^m(I\times P,N)$$

$$\longrightarrow {}_rJ^m(\mathbb{R}\times P,N)$$

is transverse to the submanifold $p_m^\infty S^*(Z,P,N)$.

LEMMA 87: Let Z' be any countable collection of invariant algebraic sets of multijets. Then any fibered concordance

$$I\times P\times D^s \xrightarrow{\;F\;} I\times N\times D^s$$

admits an arbitrarily small fibered isotopy of concordances F^u $(0\le u\le 1)$ such that $F^0 = F$ and F^1 is in general position with respect to each $Z \in Z'$.

PROOF of Lemma 87: Let $F = (h,f,p_3)$ be a fibered concordance. Apply Lemma 34 to f, taking $X = I\times P$, $Y = N$, $C = I\times\partial P$, $M = D^s$, $f_0 = f$. The conclusion is that in every neighborhood of f there is a smooth fibered homotopy.

$$I\times P\times D^s \xrightarrow{f^u} N \qquad 0\le u\le 1$$

with $f^0 = f$, f^u fixed on $I\times\partial P\times D^s$, and

$$\left. {}_r\tilde{j}^m(f^1)\right|_{(I\times \mathrm{int}P)^{(r)}\times D^s} \pitchfork p_m^\infty S^*(Z,P,N)$$

for every $Z \in Z'$ of rank r and level m. Choose the neighborhood of f small enough so that (h,f^u,p_3) is a fibered isotopy of embeddings of $I\times P$ in $I\times N$ over D^s. The proof would be done now if (h,f^u,p_3) were a fibered isotopy of concordances, because the final concordance (h,f^1,p_3) would be in general position with respect to each $Z \in Z'$. However, we have moved $0\times P\times D^s$ and we must put it back where we found it.

The map

$$P\times D^s \xrightarrow{g^u} N\times D^s$$

$$g^u(x,y) = (f^u(0,x,y),y)$$

is a smooth fibered isotopy of P in N fixing ∂P pointwise. Therefore it can be extended to a smooth fibered isotopy of N over D^s

$$N\times D^s \xrightarrow{G^u} N\times D^s$$

which fixes ∂N pointwise. Set

$$F^u(t,x,y) = (1_I\times G^u)^{-1}(h(t,x,y),f^u(t,x,y),y).$$

Then F^u is a smooth fibered isotopy of concordances of P in N over D^s, with $F^0 = F$. Moreover, the composition with $(1_I\times G^u)^{-1}$ has not affected the transversality, as one easily checks by applying Lemma 54 with $\mathcal{U}=I_{\mathbb{R}}\times G^1$. That is, F^1 is in general position with respect to each $Z \in \mathcal{Z}'$.

Finally, F^u can be made to lie in any neighborhood of F by taking f^u to lie in a suitable neighborhood of f and taking G^u to lie in a suitable neighborhood of the identity. QED.

§II.D.2. Singular sets.

§II.D.2.a. $S_\alpha(F)$

DEFINITION 88: If $F = (h,f,p_3)$ is a fibered concordance of P in N over D^s and r is the rank of $Z_\alpha \in Z$, then

$$S_\alpha(F) = ({}_r\tilde{j}^\infty(f)^{-1}S(Z_\alpha,P,N)) \cap ((I\times \mathrm{int}P)^{(r)}\times D^s)$$

and

$$S^*_\alpha(F) = ({}_r\tilde{j}^\infty(f)^{-1}S^*(Z_\alpha,P,N)) \cap ((I\times \mathrm{int}P)^{(r)}\times D^s)$$

REMARK 89. If Z_α has level m then $S^*_\alpha(F)$ is the preimage of the manifold $p^\infty_m S^*(Z\ ,P,N)$ under the map (173). It is tempting to conclude that if F is in general position with respect to Z_α then $S^*_\alpha(F)$ is a submanifold of $(I\times \mathrm{int}P)^{(r)}\times D^s$. Unfortunately the situation is not quite that simple, since the restriction of (173) to the boundary or corner set of $(I\times \mathrm{int}P)^{(r)}\times D^s$ need not be transverse. However, one can do the following: Extend f to get a smooth map

$$\hat{f} : \mathcal{U} \longrightarrow N$$

where U is an open neighborhood of $I\times P\times D^s$ in $\mathbb{R}\times P\times\mathbb{R}^s$. General position for F (w.r.t. Z) implies that

$$ {}_r\tilde{j}^m(f\big|_V) \pitchfork p_m^\infty S^*(Z_\alpha,P,N), $$

where V is some open neighborhood of $I\times\mathrm{int}P\times D^s$ in $U\cap(\mathbb{R}\times\mathrm{int}P\times\mathbb{R}^s)$. Set

$$ \hat{S}^*_\alpha(F) = {}_r\tilde{j}^m(f\big|_V)^{-1}p_m^\infty S^*(Z_\alpha,P,N) . $$

Then $\hat{S}^*_\alpha(F)$ is a submanifold of $(\mathbb{R}\times\mathrm{int}P)^{(r)}\times\mathbb{R}^s$ of dimension $s-c(Z_\alpha)$, and

$$ S^*_\alpha(F) = \hat{S}^*_\alpha(F)\cap(I\times\mathrm{int}P)^{(r)}\times D^s . $$

These comments will be useful more than once.

The rest of §II.D. is concerned with exploring relationships between the various sets $S_\alpha(F)$ (for fixed F and varying α). In particular a subset $W_\alpha(F)\subset S^*_\alpha(F)$ will be defined (Def. 94) which is, roughly speaking, the set of all points in $S_\alpha(F)$ which do not "come from" any $S_\beta(F)$, $\beta>\alpha$, by any morphism $Z_\alpha\longrightarrow Z_\beta$.

LEMMA 90: Let $Z_\beta = D_\phi(Z_\alpha)$, with $r=\mathrm{rank}(Z_\alpha)$ and $r'=\mathrm{rank}(Z_\beta)$. Let $F = (h,f,p_3)$ be a fibered concordance of P in N over D^s. If $(x_1,\cdots x_{r'},y) \in (I\times P)^{(r')}\times D^s$ is such that $(x_{\phi(1)},\cdots x_{\phi(r)},y)$ is a limit point of ${}_r\tilde{j}^\infty(f)^{-1}S(Z_\alpha,P,N)$ in $(I\times P)^r\times D^s$, then

$$(x_1,\cdots x_{r'},y) \in {}_{r'}\tilde{j}^\infty(f)^{-1}S(Z_\beta,P,N) .$$

PROOF of Lemma 90: This is an easy consequence of Lemma 61. (Just choose coordinates in P near the set $\{p_2x_1,\cdots p_2x_{r'}\}$ and in N near the set $f(x_1,\cdots f(x_{r'})$.)

QED

NOTATION: For any $r\geq 1$ and $1\leq j\leq r$ and $s\geq 0$ π_j is the projection

$$(I\times P)^r\times D^s \longrightarrow I\times P\times D^s$$

$$(x_1,\cdots x_r,y) \longmapsto (x_j,y) .$$

LEMMA 91: Let $F = (h,f,p_3)$ be a fibered concordance of P in N over D^s, and let $x \in {}_r\tilde{j}^\infty(f)^{-1}S(Z_\alpha,P,N)$. Then for any $i,j \in I(Z_\alpha)$,

$$i \sim_{Z_\alpha} j \quad \Rightarrow \quad p_{2,3}\pi_i x = p_{2,3}\pi_j x$$

$$i \simeq_{Z_\alpha} j \quad \Longrightarrow \quad p_{2,3}F\pi_i x = p_{2,3}F\pi_j x$$

$$i \in \Delta_{Z_\alpha} \quad \Rightarrow \quad \ker(D(p_{2,3} \circ F))(\pi_i x) \neq 0 \ .$$

(Here $p_{2,3}$ is the projection $I \times P \times D^s \longrightarrow P \times D^s$ or $I \times N \times D^s \longrightarrow N \times D^s$.)

PROOF: Easy. Choose a coordinate patch in P which includes $\{p_2\pi_1 x, \cdots p_2\pi_r x\}$, and a coordinate patch in N which includes $\{f\pi_1 x, \cdots f\pi_r x\}$, and chase the definitions.

QED.

LEMMA 92: Let $F = (h, f, p_3)$ be a fibered concordance of P in N over D^s, and let $x \in {}_r\tilde{J}^\infty(f)^{-1}S(Z_\alpha, P, N)$. If for some $j \in I(Z_\alpha)$ $\pi_j(x) \in I \times \mathrm{int}P \times D^s$, then $x \in S_\alpha(F)$.

PROOF: The conclusion just means that $\pi_i(x) \in I \times \mathrm{int}P \times D^s$ for <u>all</u> $i \in I(Z_\alpha)$. Set

$$\Sigma = \left\{ i \in I(Z_\alpha) \;\middle|\; \pi_i(x) \in I \times \mathrm{int}P \times D^s \right\}.$$

Lemma 91 implies that Σ is a union of $\sim_{Z_\alpha}$-classes, since

$$k \sim_{Z_\alpha} i \in \Sigma \Rightarrow p_{2,3}\pi_k x = p_{2,3}\pi_i x \in \mathrm{int}P\times D^s$$

$$\Longrightarrow \quad k \in \Sigma ,$$

and also a union of $\simeq_{Z_\alpha}$-classes, since

$$k \simeq_{Z_\alpha} i \in \Sigma \Rightarrow p_{2,3}F\pi_k x = p_{2,3}F\pi_i x \in \mathrm{int}N\times D^s$$

$$\Longrightarrow \quad k \in \Sigma .$$

By Lemma 69 (Statement (149)) it follows that either $\Sigma = \emptyset$ or $\Sigma = I(Z_\alpha)$. But $j \in \Sigma$. QED

LEMMA 93: Let F be a fibered concordance of P in N over D^s. Let $\beta \geq 0$ and $j \in I(Z_\beta)$. Then

$$\overline{\pi_j S_\beta(F)} \subset \pi_j S_\beta(F) \cup \bigcup_{Z_{\beta'} = D_\phi(Z_\beta)} \pi_{\phi(j)} S_{\beta'}(F) .$$

(The bar denotes closure in $I \times \mathrm{int}P\times D^s$.)

PROOF: This is an easy consequence of Lemmas 90 and 92. Any point in the LHS has the form $\pi_j(x)$ where $x \in (I\times P)^r \times D^s$ is a limit point of $S_\beta(F)$. (Here r is the rank of Z_β.) If $x \in (I\times P)^{(r)}\times D^s$, then $x \in {}_r\tilde{j}^\infty(f)^{-1}S(Z_\beta,P,N)$ because this latter set is closed in $(I\times P)^{(r)}\times D^s$ and contains $S_\beta(F)$. If not, then Lemma 90 implies that for some ϕ and $Z_{\beta'} = D_\phi(Z_\beta)$, $x \in \phi^*({}_{r'}\tilde{j}^\infty(f)^{-1}S(Z_{\beta'},P,N))$.

Now apply Lemma 92. In the first case $\pi_j(x) \in I\times \mathrm{int}\,P\times D^s$ implies $x \in S_\beta(F)$, which implies $\pi_j(x) \in$ RHS. In the second case, write $x=\phi^*y$. Then $\pi_{\phi(j)}(y) = \pi_j(x) \in I\times \mathrm{int} P\times D^s$ implies $y \in S_{\beta'}(F)$, which implies $\pi_j(x) = \pi_{\phi(j)}(y) \in$ RHS. QED

§II.D.2.b. $\underline{W_\alpha(F)}$.

DEFINITION 94. Let F be a fibered concordance of P in N over D^s. Let $\alpha\geq 0$. Then

$$Y_\alpha(F) = \bigcup_{\beta\geq\alpha}\ \bigcup_{j\in I(Z_\beta)} \pi_j S_\beta(F) \subset I\times \mathrm{int}\ P\times D^s$$

$$W_\alpha(F) = \left\{ x \in S_\alpha(F) \;\middle|\; \text{for all } \beta>\alpha,\ i \in I(Z_\alpha), \text{ and } j \in I(Z_\beta),\ \pi_i(x) \notin \pi_j S_\beta(F) \right\}$$

$$= \left\{ x \in S_\alpha(F) \;\middle|\; \text{for all } \; i \in I(Z_\alpha), \pi_i(x) \notin Y_{\alpha+1}(F) \right\} .$$

The rest of Chapter II is mainly concerned with seeing what is implied about the sets $W_\alpha(F)$ by the four operations.

§II.D.2.b.i. Operation A.

LEMMA 95: $W_\alpha(F) \subset S^*_\alpha(F)$.

PROOF: Choose β such that $Z_\beta = A(Z_\alpha)$. Thus $S^*_\alpha(F) = S_\alpha(F) - S_\beta(F)$. $\alpha < \beta$ by (172). Therefore by the definition of $W_\alpha(F)$

$$\pi_1 W_\alpha(F) \cap \pi_1 S_\beta(F) = \emptyset .$$

So

$$W_\alpha(F) \cap S_\beta(F) = \emptyset,$$

and $W_\alpha(F) \subset S^*_\alpha(F)$. QED.

§II.D.2.b.ii. Operation D.

LEMMA 96: If F is in general position with respect to Z then $Y_\alpha(F)$ is closed in $I\times\text{int }P\times D^s$.

PROOF: The general-position hypothesis implies that $S_\beta(F)$ is empty whenever $c(Z_\beta)>s$. Therefore, by Lemma 81, the definition of $Y_\alpha(F)$ really expresses it as a finite union. But by Lemma 93 the closure of $\pi_j S_\beta(F)$ in $I\times\text{int }P\times D^s$ is always contained in $Y_{\alpha+1}$ (for any $\beta>\alpha$ and $j\in I(Z_\beta)$). Thus $Y_{\alpha+1}(F)$ is closed. QED

LEMMA 97: If F is in general position with respect to Z then $W_\alpha(F)$ is open in $S^*_\alpha(F)$.

PROOF: Equivalently, $W_\alpha(F)$ is open in $S_\alpha(F)$. But this is immediate from Lemma 96. QED

§II.D.2.b.iii. Operation B.

LEMMA 98: Let

$$I\times P\times D^s \xrightarrow{F} I\times N\times D^s$$

be a fibered concordance. Let $\alpha, \alpha' \geq 0$. Let x and x' be points

$$x \in S_{\alpha}(F) \ , \ x' \in S_{\alpha'}(F) \ .$$

Let $i \in I(Z_{\alpha})$, $i' \in I(Z_{\alpha'})$. If

(174) $$p_{2,3}\pi_i x = p_{2,3}\pi_{i'} x'$$

then there exist an integer $\alpha'' \geq 0$, morphisms

$$Z_{\alpha} \xrightarrow{\ \phi\ } Z_{\alpha''} \xleftarrow{\ \phi'\ } Z_{\alpha'} \ ,$$

and a point

(175) $$x'' \in S_{\alpha''}(F)$$

such that

(176) $$\phi^* x'' = x$$

(177) $$\phi'^* x'' = x' \quad .$$

Moreover if $x \in W_{\alpha}(F)$ (resp. $x' \in W_{\alpha'}(F)$) then ϕ (resp. ϕ') is an isomorphism.

PROOF: Consider the finite set of points

$$\{\pi_1 x, \cdots \pi_r x, \pi_1 x', \cdots \pi_{r'} x'\} \subset I\times P\times D^s \ ,$$

where r is the rank of Z_α and r' is the rank of $Z_{\alpha'}$. Let r'' be the number of points in this set, and list them in some order:

$$y_1, \cdots y_{r''} \ .$$

Let ϕ and ϕ' be the unique maps of sets

$$I(Z_\alpha) \xrightarrow{\phi} \{1, \cdots r''\}$$

$$I(Z_{\alpha'}) \xrightarrow{\phi'} \{1, \cdots r''\}$$

such that

$$y_{\phi(j)} = \pi_j x \quad \text{for all} \quad j \in I(Z_\alpha)$$

$$y_{\phi'(j')} = \pi_{j'} x' \quad \text{for all} \quad j' \in I(Z_{\alpha'}) \ .$$

Necessarily ϕ and ϕ' are injections and the union of their images is $\{1, \cdots r''\}$. Thus it makes sense to define

$$B_{\phi,\phi',i,i'}(Z_\alpha, Z_{\alpha'}) = Z_{\alpha''} \ ,$$

an element of Z with rank r''. ϕ and ϕ' are then morphisms $Z_\alpha \longrightarrow Z_{\alpha''}$ and $Z_{\alpha'} \longrightarrow Z_{\alpha''}$ (by the proof of Lemma 74). Let x'' be the unique point

$$x'' \in (I\times \text{int } P)^{(r'')}\times D^s$$

such that

$$\pi_{j''}x'' = y_{j''} \quad \text{for all} \quad j'' \in I(Z_{\alpha''}).$$

Then it is not hard to verify (175), (176), and (177). (For (175) choose local coordinates and use (174).)

Finally, suppose $x \in W_\alpha(F)$. Then since

$$\pi_i x = \pi_{\phi(i)}x'' \in \pi_{\phi(i)}S_{\alpha''}(F) \ ,$$

the definition of $W_\alpha(F)$ implies that $\alpha''\leq\alpha$. Unless ϕ is an isomorphism, we must have

$$[Z_\alpha] \lneq [Z_{\alpha''}] \ ,$$

so that by (171) $\alpha<\alpha'$, a contradiction. Likewise $x' \in W_{\alpha'}(F) \Rightarrow \phi'$ is an isomorphism. QED.

With Lemma 98 at our disposal we can prove:

LEMMA 99: Let F be a fibered concordance of P in N over D^s. Let $\alpha\geq 0$, $\alpha'\geq 0$, $\alpha\neq\alpha'$, $i\in I(Z_\alpha)$, $i'\in I(Z_\alpha)$. Then

$$p_{2,3}\pi_i W_\alpha(F) \cap p_{2,3}\pi_{i'} W_{\alpha'}(F) = \emptyset$$

PROOF: Suppose

$$p_{2,3}\pi_i x = p_{2,3}\pi_{i'} x'$$

$$x \in W_\alpha(F),\ x' \in W_{\alpha'}(F)$$

Lemma 98 applies and gives isomorphisms

$$Z_\alpha \xrightarrow{\ \phi\ } Z_{\alpha''} \xleftarrow{\ \phi'\ } Z_{\alpha'}\ .$$

$\pi_i x = \pi_{\phi(i)} x'' = \pi_{\phi'^{-1}(\phi(i)} x' \in \pi_{\phi'^{-1}(\phi(i))} S_{\alpha'}(F)$, so by the definition of $W_\alpha(F)$ $\alpha'\leq\alpha$. Likewise $\alpha\leq\alpha'$. Thus $\alpha=\alpha'$, a contradiction. QED.

Another consequence of Lemma 98, or rather of its proof, is:

LEMMA 100: The implications ($\Rightarrow$) of Lemma 91 become reversible ($\Longleftrightarrow$) if $x \in W_\alpha(F)$.

PROOF: First suppose that $x \in W_\alpha(F)$ and $p_{2,3}\pi_i x = p_{2,3}\pi_j x$. Apply Lemma 98 with $\alpha=\alpha'$, $i'=j$, $x=x'$. The set $Z_{\alpha''}$ which results can be taken to be $B_{id,id,i,j}(Z_\alpha,Z_\alpha)$, and the morphism $\phi\colon Z_\alpha \longrightarrow Z_{\alpha''}$ is then given by the identity map $I(Z_\alpha) \xrightarrow{=} I(Z_{\alpha''})$. Since ϕ is an isomorphism (by Lemma 98) we have $Z_\alpha = Z_{\alpha''}$ and thus by Def. 68 $i \sim_{Z_\alpha} j$.

Next suppose that $x \in W_\alpha(F)$ and $p_{2,3}F\pi_i x = p_{2,3}F\pi_j x$. We may assume $i \neq j$. Define

$$x' \in (I\times P)^{(2)}\times D^s$$

by

$$\pi_1 x' = \pi_i x \ , \ \pi_2 x' = \pi_j x \ .$$

Then $x' \in S_0(F)$. Apply Lemma 98 with $\alpha'=0$, $i'=1$. Again the ϕ of Lemma 98 can be taken to be given by the identity $I(Z_\alpha) \xrightarrow{=} I(Z_{\alpha''})$. We conclude that
$Z_\alpha = Z_{\alpha''} = B_{id,\phi',i,1}(Z_\alpha,Z_0)$, so that $i \simeq_{Z_\alpha} j$ by Def. 68.

Finally, assume $x \in W_\alpha(F)$ and

$$\ker(D(p_{2,3} \circ F))(\pi_i x) \neq 0 .$$

Set $x' = \pi_i x$, $\alpha'=1$, $i'=1$; apply Lemma 98 and argue as in the previous case to conclude that $i \in \Delta_{Z_\alpha}$. QED.

A third consequence of Lemma 98 which we will eventually need is:

LEMMA 101: Let F be a fibered concordance, $\alpha \geq 0$. Then

$$(178)\quad W_\alpha(F) = \left\{ x \in S_\alpha(F) \;\middle|\; \text{for all } \beta > \alpha \text{ and all morphisms } \phi: Z_\alpha \to Z_\beta\ x \notin \phi^* S_\beta(F) \right\}$$

$$(179)\quad W_\alpha(F) = \left\{ x \in S_\alpha(F) \;\middle|\; \text{for all } i \in I(Z_\alpha)\ p_{2,3}\pi_i(x) \notin p_{2,3} Y_{\alpha+1} \right\} .$$

($Y_{\alpha+1}$ was defined in Def. 94, §II.D.2.b.)

PROOF: Clearly

$$[\text{R.H.S. of } (179)] \subset W_\alpha(F) \subset [\text{R.H.S. of } (178)]$$

Now suppose $x \notin$ [R.H.S. of (179)]. Then for some $\alpha'>\alpha$, $i \in I(Z_\alpha)$, $i' \in I(Z_{\alpha'})$, and $x' \in S_{\alpha'}(F)$, we have

$$p_{2,3}\pi_i(x) = p_{2,3}\pi_{i'}(x').$$

Apply Lemma 98 to get α'', $x'' \in S_{\alpha''}(F)$, and morphisms

$$Z_\alpha \xrightarrow{\phi} Z_{\alpha''} \xleftarrow{\phi'} Z_{\alpha'}$$

such that

$$\phi^* x'' = x \ , \ \phi'^* x'' = x' \ .$$

If ϕ is not an isomorphism, then $\alpha''>\alpha$ by (171) so that $x \notin$ RHS of (178). If ϕ is an isomorphism then the morphism $\phi'\cdot\phi^{-1}$ (where $\alpha'>\alpha$ by assumption) shows that $x \notin$ RHS of (178). QED.

§II.D.2.b.iv. Operation C.

LEMMA 102: If F is in general position with respect to Z_α then for each $i \in I(Z_\alpha)$ the map

$$\hat{S}^*_\alpha(F) \xrightarrow{p_{2,3}\cdot\pi_i} P\times D^s$$

is an immersion at all points in $W_\alpha(F)$. ($\hat{S}^*_\alpha(F)$ was defined in Remark 89.)

PROOF: Assume that $x \in W_\alpha(F)$ and that the kernel of

$$D(p_{2,3}\circ\pi_i\Big|_{\hat{S}^*_\alpha(F)})(x)$$

contains a vector

$$v \in T_x((I\times P)^{(r)}\times D^s)$$

different from zero. The fact that v is in the kernel means that the vector

$$(D\pi_i)\cdot v \in T_{\pi_i(x)}(I\times P\times D^s)$$

is "vertical" (Def. 26, §I.C). The fact that $v \neq 0$ means that for some $j \in I(Z_\alpha)$

$$(D\pi_j)\cdot v \neq 0 \ .$$

CLAIM 103: There is some $k \in I(Z_\alpha)$ such that

$$(D(p_{2,3}\circ F))\cdot(D\pi_k)\cdot v = D(p_{2,3}\circ F)\cdot w \tag{180}$$

for some vertical vector $w \in T_{\pi_k(x)}(I\times P\times D^s)$, and

$$(D\pi_k)\cdot v \neq 0 \ . \tag{181}$$

PROOF: Set

$$I_1 = \left\{ k \in I(Z_\alpha) \;\middle|\; (180) \text{ holds for some } w \right\}$$

$$I_2 = \left\{ k \in I(Z_\alpha) \;\middle|\; (181) \text{ holds} \right\} \ .$$

Certainly $I_1 \cup I_2 = I(Z_\alpha)$. Also $i \in I_1$ and $j \in I_2$, so I_1 and I_2 are not empty. The claim is that $I_1 \cap I_2 \neq \emptyset$.

Assume the contrary. Then by Lemma 69 part (149) there exist $i' \in I_1 - I_2$ and $j' \in I_2 - I_1$ such that

$$\text{either} \quad i' \sim_{Z_\alpha} j' \quad \text{or} \quad i' \simeq_{Z_\alpha} j' .$$

Since $i' \notin I_2$, we have

$$(D\pi_{i'}) \cdot v = 0 . \tag{182}$$

If $i' \sim_{Z_\alpha} j'$, then

$$p_{2,3} \circ \pi_{i'} \Big|_{\hat{S}^*_\alpha(F)} = p_{2,3} \circ \pi_{j'} \Big|_{\hat{S}^*_\alpha(F)} \quad \text{by Lemma 91}$$

so that

$$(Dp_{2,3}) \cdot (D\pi_{j'}) \cdot v = (Dp_{2,3}) \cdot (D\pi_{i'}) \cdot v = 0 \quad \text{by (182)},$$

i.e., $(D\pi_{j'}) \cdot v$ is vertical and $j' \in I_1$, a contradiction. If $i' \simeq_{Z_\alpha} j'$, then

$$p_{2,3} \circ F \circ \pi_{i'} \Big|_{\hat{S}^*_\alpha(F)} = p_{2,3} \circ F \circ \pi_{j'} \Big|_{\hat{S}^*_\alpha(F)} \quad \text{by Lemma 91,}$$

so that

$$D(p_{2,3}\circ F)\cdot(D\pi_{j'})\cdot v = D(p_{2,3}\circ F)\cdot(D\pi_{i'})\cdot v$$

$$= D(p_{2,3}\circ F)\cdot w \quad \text{for some vertical } w, \quad \text{since } i' \in I_1$$

i.e., $j' \in I_1$ again. This contradiction proves the Claim.

QED.

Now let k be as in the Claim. It follows by choosing local coordinates in P and N that

$$_r\tilde{j}^\infty(f)(x) \in S(C_k(Z_\alpha),P,N) .$$

Thus $x \in S_\beta(F)$, where $Z_\beta = C_k(Z_\alpha)$. But $\beta>\alpha$ by (172). Therefore $x \notin W_\alpha(F)$, a contradiction. QED.

If $i \in \Delta_{Z_\alpha}$ (the set defined in Def. 68) then the maps

$$S_\alpha(F) \xrightarrow{\pi_i} I\times P\times D^s \xrightarrow{p_{2,3}} P\times D^s$$

carry some additional structure. Namely, for each point $x \in S_\alpha(F)$ there is a unique vector

$$\xi_{\alpha,i}(x) \in T_{\pi_i(x)}(I\times P\times D^s)$$

such that

$$(DF)(\pi_i(x))\cdot\xi_{\alpha,i}(x) = \frac{\partial}{\partial t} .$$

(Lemma 100 implies that there exists non-zero $\xi \in T_{\pi_i(x)} I\times P\times D^s$ such that $DF\cdot\xi$ is a multiple of $\frac{\partial}{\partial t}$, and the fact that F is an immersion finishes the job.) Thus $\xi_{\alpha,i}$ is a vector field along the map π_i. Set

$$\eta_{\alpha,i}(x) = (Dp_{2,3})(\pi_i(x))\cdot\xi_{\alpha,i}(x) \in T_{(p_{2,3}\circ\pi_i)(x)}(P\times D^s).$$

LEMMA 104: If Z_α is good (Def. 84) and $x \in W_\alpha(F)$, then for any $i \in \Delta_{Z_\alpha}$

$$\eta_{\alpha,i}(x) \notin D(p_{2,3}\circ\pi_i)\cdot T_x\hat{S}^*_\alpha(F) .$$

PROOF of Lemma 104: Suppose that $\eta_{\alpha,i}(x) = D(p_{2,3}\circ\pi_i)\cdot v$, $v \in T_x\hat{S}^*_\alpha(F)$. Then $\xi_{\alpha,i}(x) - (D\pi_i)\cdot v$ is vertical. If $\eta_{\alpha,i}(x) \neq 0$, then Claim 103 is valid (with $k=i$) and we get a contradiction as in the proof of Lemma 102. Suppose that $\eta_{\alpha,i}(x) = 0$. Thus $\xi_{\alpha,i}(x)$ is vertical and $\pi_i(x) \in S(\hat{Z},P,N)$ (see Def. 82). Now by Lemma 101 $x \in W_\alpha(F) \Rightarrow [\hat{Z}] \leq [Z_\alpha] \Rightarrow Z_\alpha$ is bad, a contradiction.

QED.

§II.D.2.c. $W_\alpha^{(D,R)}(F)$.

Let $x \in W_\alpha(F)$. We have seen (Lemma 100) that the relation $\sim_{Z_\alpha}$ determines when two of the points $\{\pi_1 x, \ldots \pi_r x\}$ $(r = \operatorname{rank}(Z_\alpha))$ will lie on the same "vertical line" $I \times x_0 \times y_0 \subset I \times \operatorname{int} P \times D^s$, and that the relation $\simeq_{Z_\alpha}$ determines when two of the points $\{F\pi_1 x, \ldots F\pi_r x\}$ will lie on the same "vertical line" $I \times x_0 \times y_0 \subset I \times \operatorname{int} N \times D^s$. We will eventually have to pay attention to the ordering of points on these lines. This is the reason for the next two definitions.

DEFINITION 105: Let $\alpha \geq 0$. A pair (D,R) of binary relations on $I(Z_\alpha)$ is called <u>admissible</u> for Z_α if D and R are both strict partial orderings and if for all i and j in $I(Z_\alpha)$

$$i \sim_{Z_\alpha} j \iff \begin{cases} iDj & \text{or} \\ i=j & \text{or} \\ jDi & \end{cases}$$

$$i \simeq_{Z_\alpha} j \iff \begin{cases} iRj & \text{or} \\ i=j & \text{or} \\ jRi & \end{cases}$$

DEFINITION 106: Let $\alpha \geq 0$, let (D,R) be an admissible pair of relations for Z_α, and let F be a fibered concordance. Then

$$W_\alpha^{(D,R)}(F) = \left\{ x \in W_\alpha(F) \;\middle|\; \text{for all } i,j \in I(Z_\alpha) \right.$$

$$\left(iDj \iff \pi_i x \text{ is below } \pi_j x \text{ in } I \times P \times D^s \right) \text{ and}$$

$$\left. \left(iRj \iff F\pi_i x \text{ is below } F\pi_j x \text{ in } I \times N \times D^s \right) \right\}$$

("Below" is defined in Def.26, §I.C.) Lemma 100 easily implies that $W_\alpha(F)$ is the disjoint union of all of the $W_\alpha^{(D,R)}(F)$.

The rest of §II.D. is concerned with proving some geometrical properties of the sets $W_\alpha^{(D,R)}(F)$.

LEMMA 107: $W_\alpha^{(D,R)}(F)$ is open in $W_\alpha(F)$.

PROOF:

$$W_\alpha^{(D,R)}(F) = \left\{ x \in W_\alpha(F) \;\middle|\; \text{for all } i \text{ and } j \text{ in } I(Z_\alpha) \right.$$

$$iDj \Rightarrow \pi_j x \text{ is not below } \pi_i x$$

$$\left. \text{and } iRj \Rightarrow F\pi_j x \text{ is not below } F\pi_i x \right\}$$

QED.

LEMMA 108: If Z_α is good and (D,R) is an admissible pair of relations for Z_α, then the only automorphism of Z_α which preserves D and R is the identity.

PROOF: Suppose $\phi: I(Z_\alpha) \longrightarrow I(Z_\alpha)$ is such an automorphism. Thus ϕ preserves the relations $\sim_{Z_\alpha}$ and $\simeq_{Z_\alpha}$ on $I(Z_\alpha)$, just because it is an automorphism. For this proof let us omit the subscripts and write simply $\sim, \simeq$.

If ϕ takes some $\sim$-class to itself then ϕ must fix every member of the class, because ϕ preserves D and D restricts to a (strict) linear ordering on the class. The same holds for any $\simeq$-class, using R instead of D. Therefore the set of all members of $I(Z_\alpha)$ fixed by ϕ is both a union of $\sim$-classes and a union of $\simeq$-classes. By Lemma 69 part (149) ϕ is the identity if it fixes even one element of $I(Z_\alpha)$. In fact, it suffices to show that ϕ fixes either an element of $I(Z_\alpha)$, or a $\sim$-class, or a $\simeq$-class.

Make a graph (one-dimensional CW complex) Γ as follows: Take vertices corresponding to the $\sim$-classes and the $\simeq$-classes, and edges corresponding to the elements of $I(Z_\alpha)$. Attach the edge which corresponds to $i \in I(Z_\alpha)$ to the two vertices which correspond to the $\sim$-class and $\simeq$-class which contain i. Γ is connected, by (149) of

Lemma 69. In fact, it is simply-connected: if it were not, then there would be a non-self-intersecting cellular loop in Γ, and hence distinct elements $i_1,\dots i_{2k} \in I(Z_\alpha)$ (for some $k>0$) such that

$$i_1 \simeq i_2 \sim i_3 \simeq \dots \simeq i_{2m} \sim i_1 .$$

But then the map

$$\psi : \{1,\dots 2k\} \longrightarrow I(Z_\alpha)$$

would be a morphism from $\hat{Z}_k$ to Z_α, contradicting goodness. Thus Γ is contractible, and the Lefschetz Fixed Point Theorem says that the obvious action of ϕ on Γ has a fixed edge or vertex. QED.

LEMMA 109: Let F be a fibered concordance over D^s, let Z_α be good, let $i,i' \in I(Z_\alpha)$, and let (D,R) and (D',R') be admissible pairs for Z_α. Then

(183) $$p_{2,3}\circ\pi_i : W_\alpha^{(D,R)}(F) \longrightarrow P\times D^s \quad \text{is injective,}$$

and

(184) $p_{2,3}\pi_i W^{(D,R)}(F)$ and $p_{2,3}\pi_{i'} W^{(D',R')}(F)$ are either equal or disjoint, according to whether or not there is an automorphism of Z_α sending the $\sim_{Z_\alpha}$-class of i to that of i' and sending D to D' and R to R'.

PROOF: The proof is based on Lemmas 98 and 108.

For (183) suppose

$$p_{2,3}\pi_i x = p_{2,3}\pi_i x'$$

$$x, x' \in W_\alpha^{(D,R)}(F) .$$

Apply Lemma 98 with $\alpha'=\alpha$, $i'=i$. This yields isomorphisms

$$Z_\alpha \xrightarrow{\ \phi\ } Z_{\alpha''} \xleftarrow{\ \phi'\ } Z_\alpha$$

and therefore an automorphism $\phi'^{-1}\cdot\phi$ such that

$$(\phi'^{-1}\circ\phi)^* x = x' .$$

Because x and x' are both in $W_\alpha^{(D,R)}(F)$, $\phi'^{-1}\circ\phi$ preserves D and R. Therefore, by Lemma 108, $\phi'=\phi$ and $x' = x$.

For (184) suppose

$$p_{2,3}\pi_i x = p_{2,3}\pi_{i'} x'$$

$$x \in W_\alpha^{(D,R)}(F) \quad , \quad x' \in W_\alpha^{(D',R')}$$

Apply Lemma 98 with $\alpha'=\alpha$. This yields isomorphisms

$$Z_\alpha \xrightarrow{\ \phi\ } Z_{\alpha''} \xleftarrow{\ \phi'\ } Z_\alpha$$

and therefore an automorphism $\phi'^{-1}\circ\phi$. $\phi'^{-1}\circ\phi$ takes (D,R) to (D',R'), and by Lemma 100 $\phi'^{-1}(\phi(i)) \sim_{Z_\alpha} j$. QED.

The final result of Chapter II (Lemma 110 below) ties together a number of the lemmas of the last few subsections:

LEMMA 110: Let F be a fibered concordance of P in N over D^s. Let Z_α be good and $i \in I(Z_\alpha)$, and let (D,R) be an admissible pair for Z_α. Then $p_{2,3}\pi_i : W_\alpha^{(D,R)}(F) \longrightarrow P\times D^s$ is a (topological) embedding, and its image is a closed subset of $(\text{int } P)\times D^s - p_{2,3}Y_{\alpha+1}$

PROOF: Certainly the image is contained in $\mathrm{int}P\times D^s$, because $W_\alpha^{(D,R)}(F) \subset S_\alpha(F) \subset (I\times \mathrm{int}\ P)^{(r)}\times D^s$. Also the image is disjoint from $Y_{\alpha+1}$ by (179) of Lemma 101.

It remains to show that if x^ν is a sequence in $W_\alpha^{(D,R)}(F)$ such that $p_{2,3}\pi_i x^\nu$ converges in

$$(\mathrm{int}P)\times D^s - p_{2,3}Y_{\alpha+1}$$

then x^ν converges in $W_\alpha^{(D,R)}(F)$. Certainly x^ν has at least one limit point x in the compact set $(I\times P)^r\times D^s$. I.e., some subsequence converges to some x. Moreover $p_{2,3}\pi_i x = \lim\limits_{\nu} p_{2,3}\pi_i x^\nu \in \mathrm{int}P\times D^s - p_{2,3}Y_{\alpha+1}$. The argument used in the proof of Lemma 93 shows that $x \in S_\alpha(F)$. An application of Lemma 98 shows that in fact $x \in W_\alpha(F)$. The fact that $W_\alpha^{(D,R)}(F)$ is closed in $W_\alpha(F)$ ($W_\alpha(F)$ is the disjoint union of open subsets $W_\alpha^{(D',R')}(F)$) implies that $x \in W_\alpha^{(D,R)}(F)$.

Thus every limit point of x^ν is in $W_\alpha^{(D,R)}(F)$. But by (183) x^ν can have at most one limit point in $W_\alpha^{(D,R)}(F)$. It follows that x^ν converges in $W_\alpha^{(D,R)}(F)$. QED.

Chapter III. Proof of Theorem D.

In this Chapter we will use all of the tools developed in Chapters I and II to prove the main result of this thesis.

THEOREM D. ("Multiple Disjunction Lemma") Let N^n be a smooth compact manifold. Let P^p and $\left\{Q_j^{q_j}\right\}_{j=1}^{a}$ $(a\geq 1)$ be disjoint compact proper submanifolds, and assume $n-p\geq 3$ and $n-q_j\geq 3$ for all j. Then the $(a+1)$-ad homotopy groups

$$\pi_i\left(C(P,N)\ ;\ \left\{C(P,N-Q_j)\right\}_{j=1}^{a}\right)$$

are trivial for $i\leq n-p-3+\sum_{j=1}^{a}(n-q_j-1)$.

§III. A. The Structure of the Proof.

Let N^n, P^p, a, and $\left\{Q_j\right\}_{j=1}^{a}$ be as in the hypothesis of Theorem D (and if $p=0$ assume that P is a point — see the footnote in §II.A.2). Apply Lemma 18 (§I.B.4) with

$$\underline{X}=\left(C(P,N)\ ;\ \left\{C(P,N-Q_j)\right\}_{j=1}^{a}\right)$$

$$k_T=n-p-3+\sum_{j\in T}(n-q_j-1)\ .$$

The conclusion of Theorem D will follow if we can show that $\underline{X}$ is k-connected, i.e., that for each nonempty set $T \subset \{1,\cdots a\}$ we have

$$(185)_T \quad \begin{cases} \text{the pair } \left(C(P,N-\bigcup_{k\notin T} Q_k), \bigcup_{j\in T} C(P,N-\bigcup_{k\notin T} Q_k - Q_j)\right) \\ \text{is } [n-p-3+\sum_{j\in T}(n-q_j-1)]\text{-connected .} \end{cases}$$

In fact if we can prove

$$(186) \quad \begin{cases} \text{the pair } \left(C(P,N), \bigcup_{j=1}^{a} C(P,N-Q_j)\right) \\ \text{is } [n-p-3+\sum_{j=1}^{a}(n-q_j-1)]\text{-connected,} \end{cases}$$

we will be done because (186) applied to $N-\bigcup_{k\notin T} Q_k$ gives $(185)_T$.

Let $0 \le s \le n-p-3+\sum_{j=1}^{a}(n-q_j-1)$. As it stands (186) is a statement about continuous maps of pairs

$$(D^s,\partial D^s) \longrightarrow (C(P,N), \bigcup_{j=1}^{a} C(P,N-Q_j)) .$$

It is equivalent to a statement about "smooth" maps. (If $C(P,N)$ were a finite-dimensional manifold then this would be true by [Hi] Theorem 3.3 p.57. There are at least two ways to bridge this gap: either reprove the theorem for maps into an infinite-dimensional smooth manifold, or else consider maps $I\times P\times D^s \longrightarrow I\times N$ and observe that the standard technique for approximating continuous maps by smooth ones preserves "smoothness in the $(I\times P)$-direction.") A "smooth" map of the pairs above means a fibered concordance

$$I\times P\times D^s \xrightarrow{F} I\times N\times D^s$$

such that

$$(187) \qquad \forall_{y\in\partial D^s}\ \exists_j\ 1\le j\le a\quad F(I\times P\times y)\cap(I\times Q_j\times D^s)=\phi$$

Consider isotopy classes of such F's, where the isotopies which we consider are fibered isotopies of concordances F^u (see §I.A.1 for definitions) such that for each u F^u satisfies (187). What we have to prove is that <u>every isotopy class of</u> F <u>satisfying</u> (187) <u>has a representative</u> F <u>satisfying</u>

$$(188) \qquad \forall_{y\in D^s}\ \exists_j\ 1\le j\le a\ F(I\times P\times y)\cap(I\times Q_j\times D^s)=\phi\ .$$

The proof that every class has such a representative will use the following inductive hypothesis, which says that in some sense none of the sets $S_\beta(F)$, $\beta\ge\alpha$, is tangled up with the Q_j's .

$$(189)_\alpha \quad \left\{ \begin{array}{l} \text{Every isotopy class (of } F \text{ satisfying (187))} \\ \text{has a representative } (h,f,p_3) = F \text{ in general} \\ \text{position with respect to } Z, \text{ such that for} \\ \text{some open cover } \left\{ \mathcal{O}_j \right\}_{j=1}^{a} \text{ of } D^s, \text{ for all} \\ j\,(1 \le j \le a): \\ \quad f(I \times (P \times \overline{\mathcal{O}}_j \cap p_{2,3} Y_\alpha)) \cap Q_j = \phi \; . \end{array} \right.$$

The overall outline of the proof is this: by (172) there is some $\alpha_0 \geq 0$ such that

$$(190) \qquad \forall_{\alpha \geq 0} \; c(Z_\alpha) \geq n-p-2 \iff \alpha \geq \alpha_0 \; .$$

In §B we will verify the statement $(189)_{\alpha_0}$ by a general position argument, using the " $\Leftarrow$ " part of (190). In §C we will prove the implication $(189)_{\alpha+1} \Rightarrow (189)_\alpha$ for any $\alpha < \alpha_0$ by using the " $\Rightarrow$ " part of (190) to construct a smooth sunny collapse and hence a fibered isotopy. The isotopy will "disentangle" one more set $S_\alpha(F)$ from the Q_j's. In §D we will use $(189)_0$ to complete the proof, using one more sunny collapse. The hard work is in §C.

§III.B. Proof of $(189)_{\alpha_0}$.

The idea of the proof is this: Find a representative F in the isotopy class which is in general position with respect to Z and which has certain other general position properties (to be specified later). Set

$$V_j = \left\{ y \in D^s \;\middle|\; \forall \alpha \geq \alpha_0 \ \forall x \in S_\alpha(F) \ \forall t \in I \ \forall i \in I(Z_\alpha) \right.$$
$$\left. p_3(x) = y \ \Rightarrow \ F(t, p_2 \pi_i x, y) \notin I \times Q_j \times D^s \right\}$$

for $1 \leq j \leq a$. If $\left\{V_j\right\}_{j=1}^{a}$ is an open cover of D^s , then we can choose another open cover $\left\{\mathcal{O}_j\right\}_{j=1}^{a}$ such that $\overline{\mathcal{O}}_j \subset V_j$ for all j. This $\left\{\mathcal{O}_j\right\}$ will meet the requirements of $(189)_{\alpha_0}$

What must be true of F in order that $\left\{V_j\right\}$ be an open cover of D^s ? The openness is automatic :

CLAIM 111: If F is in general position with respect to Z then V_j is open in D^s .

PROOF: The complement $D^s - V_j$ can be written

$$p_3\left(p_{2,3}^{-1}\, p_{2,3}(Y_{\alpha_0}(F)) \cap F^{-1}(I \times Q_j \times D^s)\right) ,$$

where p_3 and $p_{2,3}$ are the projections

$$D^s \xleftarrow[p_3]{} I\times(\mathrm{int}P)\times D^s \xrightarrow[p_{2,3}]{} (\mathrm{int}P)\times D^s .$$

Now, $p_{2,3}^{-1}\, p_{2,3}\, Y_{\alpha_0}(F)$ is closed in $I\times\mathrm{int}\, P\times D^s$ by Lemma 96 and the properness* of $p_{2,3}$. Also $F^{-1}(I\times Q_j\times D^s)$ is a compact subset of $I\times\mathrm{int}\, P\times D^s$ because it is closed in $I\times P\times D^s$. Therefore D^s-V_j is compact. QED.

What would it mean if the $\{V_j\}$ failed to cover D^s? (The idea is to show that generically they do cover.) Suppose $y\in D^s$ and $y\notin V_j$ for all $j(1\le j\le a)$. This means that for each j there exist $\alpha_j\ge\alpha_0$, $x_j\in S_{\alpha_j}(F)$, $t_j\in I$, and $i_j\in I(Z_{\alpha_j})$ such that

$$y = p_2(x_j)$$

$$F(t_j, p_2\pi_{i_j}x_j, y)\in I\times Q_j\times D^s . \tag{191}$$

Choose these α_j, x_j, t_j, and i_j so as to minimize the total number of distinct pairs (α_j, x_j). Then:

---

*I.e., in the sense of general topology: a map is proper if the preimage of every compact set is compact.

CLAIM 112. $\pi_i x_k = \pi_{i'} x_{k'} \Rightarrow (\alpha_k, x_k) = (\alpha_{k'}, x_{k'})$. for any k, i, k', and i' (such that $1 \le k \le a$, $i \in I(Z_{\alpha_k})$, $1 \le k' \le a$, and $i' \in I(Z_{\alpha_{k'}})$).

PROOF. Suppose $\pi_i x_k = \pi_{i'} x_{k'}$. Apply Lemma 98 with $\alpha = \alpha_k$, $\alpha' = \alpha_{k'}$, $x = x_k$, $x' = x_{k'}$. This yields α'', morphisms

$$Z_{\alpha_k} \xrightarrow{\phi} Z_{\alpha''} \xleftarrow{\phi'} Z_{\alpha_{k'}} ,$$

and $x'' \in S_{\alpha''}(F)$ such that $\phi^* x'' = x$, $\phi'^* x'' = x'$. Change the data $(\{\alpha_j\}, \{x_j\}, \{t_j\}, \{i_j\})$ by replacing

$$\alpha_k \text{ by } \alpha'' \qquad \alpha_{k'} \text{ by } \alpha''$$

$$x_k \text{ by } x'' \qquad x_{k'} \text{ by } x''$$

$$i_k \text{ by } \phi(i_k) \qquad i_{k'} \text{ by } \phi'(i_{k'}) .$$

The new data still satify the required conditions, that is, $\alpha'' \ge \alpha_0$, $x'' \in S_{\alpha''}(F)$, $y = p_2(x'')$, $F(t_k, p_2 \pi_{\phi(i_k)} x'', y) \in I \times Q_k \times D^s$, and $F(t_{k'}, p_2 \pi_{\phi'(i_{k'})} x'', y) \in I \times Q_{k'} \times D^s$. But both (α_k, x_k) and $(\alpha_{k'}, x_{k'})$ have been replaced by (α'', x''), so by minimality they must have been equal to begin with. QED.

Now enumerate the distinct pairs (α_j, x_j) :

$$(\alpha^1, x^1), \cdots (\alpha^b, x^b)$$

That is, for some surjection $\psi: \{1,\cdots a\} \longrightarrow \{1,\cdots b\}$, set

$$(\alpha_j, x_j) = (\alpha^{\psi(j)} , x^{\psi(j)}) \quad (1 \le j \le a) .$$

Let $r_k = \operatorname{rank}(I_{Z_{\alpha^k}})$ and $r = \sum_{k=1}^{b} r_k$. Set

$$x = (p_1 x^1, \cdots p_1 x^b, y) \in (I\times P)^{(r_1)} \times \ldots \times (I\times P)^{(r_b)} \times D^s \subset (I\times P)^r \times D^s .$$

Claim 112 implies that $x \in (I\times P)^{(r)} \times D^s$.

In fact we have

$$\text{(192)} \qquad x \in {}_r\tilde{j}^\infty(f)^{-1} S(\mathcal{Z},P,N) \cap (I\times \operatorname{int}P)^{(r)} \times D^s$$

for a certain IASCM $\mathcal{Z}$ (<u>not</u> an element of Z) which will now be defined. Set

$$\mathcal{Z} = \left\{ (z_1,\cdots z_r) \in {}_rJ^\infty \;\middle|\; (z_1,\cdots z_{r_1}) \in Z_{\alpha^1} , \right.$$
$$(z_{r_1+1},\cdots z_{r_1+r_2}) \in Z_{\alpha^2} ,\cdots$$
$$\left. (z_{r-r_b+1},\cdots, z_r) \in Z_{\alpha^b} \right\}$$

It is easy to see that Z is an IASCM of rank r and level $\max(\mathrm{level}(Z_{\alpha^k}))$, with

$$c(Z) = \sum_{k=1}^{b} c(Z_{\alpha^k}) \tag{193}$$

Thus $S(Z,P,N)$ is defined and (192) is clear.

Setting $i'_j = i_j + \sum_{k=1}^{\psi(j)-1} r_k$ $(1 \le j \le a)$ we also have

$$F(t_j, p_2\pi_{i'_j} x, y) \in I \times Q_j \times D^s \tag{194}$$

for all j, by (191).

We now show that if F is in suitably general position then (192) and (194) cannot happen.

First use Lemma 87, applying it to the collection of IASCM's consisting of Z and all possible sets $A^h(Z)$, where $h \ge 0$ and Z is defined in terms of some b $(1 \le b \le a)$ and $\alpha^1, \dots \alpha^b$ $(\alpha^k \ge \alpha_0)$ as above. We conclude that by a fibered isotopy of concordances F can be put in general position with respect to each $A^h(Z)$ as well as each Z_α. Also the isotopy can be made small enough so that (187) is preserved.

Consider the sets

$$\Sigma \underset{\text{def}}{=} {}_r\tilde{j}^\infty(f)^{-1}S^*(A^h(Z),P,N) \cap [(I\times \text{int}P)^{(r)}\times D^s] .$$

(Σ thus depends on a choice of $h,b,\alpha^1,\cdots\alpha^b$.) These would be manifolds if it were not for the difficulty discussed in Remark 89. We will pretend that they are manifolds. (The reader may fill in this gap by using that same Remark.)

The dimension of Σ is

$$s-c(A^h(Z)) \leq s-c(Z)$$

$$= s-\sum_{k=1}^{b} c(Z_{\alpha^k}) \quad \text{by (193)}$$

$$\leq s-(n-p-2) \quad \text{by (190)}$$

$$< \sum_{j=1}^{a} (n-q_j-1) \quad \text{by assumption.}$$

The map

$$I^a\times\Sigma \xrightarrow{T} N^a$$

$$(s_1,\cdots s_a,x) \longmapsto (f(s_1,p_2\pi_{i_1}x,p_2x) ,$$

$$\ldots f(s_a,p_2\pi_{i_a}x,p_2x))$$

ought to have image disjoint from $Q_1 \times \ldots \times Q_a$, because

$$(195) \qquad \dim(I^a \times \Sigma) < \sum_{j=1}^{a} (n-q_j) = \operatorname{codim}(Q_1 \times \ldots \times Q_a, N^a) .$$

To replace "ought to have" by "has" we move the Q_j's . First find an isotopy of N fixing $P \cup Q_2 \cup \ldots \cup Q_a \cup \partial N$ pointwise and taking Q_1 to some Q_1' such that the composition

$$I^a \times \Sigma \xrightarrow{T} N^a \xrightarrow{p_1} N$$

is transverse to Q_1' . Then similarly move Q_2 , fixing $P \cup Q_1' \cup Q_3 \cup \cdots \cup Q_a \cup \partial N$ pointwise, to some Q_2' such that

$$T^{-1}p_1^{-1}Q_1' \hookrightarrow I^a \times \Sigma \xrightarrow{T} N^a \xrightarrow{p_2} N$$

is transverse to Q_2' (and therefore

$$I^a \times \Sigma \xrightarrow{T} N^a \xrightarrow{p_{1,2}} N^2$$

is transverse to $Q_1' \times Q_2'$). Continue in this way until at last T is transverse to $Q_1' \times \cdots \times Q_a'$. By (195), the image of T is then disjoint from $Q_1' \times \ldots \times Q_a'$. Finally, if G^u $(0 \le u \le 1)$ is an isotopy of N rel ∂N which takes Q_j to Q_j' for all j, then the fibered isotopy of concordances

$$F^u = (1_I \times G^u \times 1_{D^s})^{-1} \circ F$$

does the required job. That is, F^1 is in general position with respect to Z (use Lemma 54) and F^u satisfies (187) for all u (if G^u is close enough to the identity), and for F^1 the image of T is disjoint from $Q_1 \times \ldots \times Q_a$. Of course, the construction of G^u must be carried out in such a way that all of this works for all Σ and all T, i.e., for all choices of b $(1 \le b \le a)$, $\alpha^1, \ldots \alpha^b$ $(\alpha^k \ge \alpha_0)$, $h \ge 0$, $i_1', \ldots i_a'$ $(1 \le i_j' \le \sum_{k=1}^{b} \text{rank}(Z_\alpha)$. Since this is a countable list of conditions, there is no problem.

Now no $x \in (I \times P)^{(r)} \times D^s$ can satisfy (192) and (194). For if x satisfies (192) for some one of our sets Z then, because F is in general position with respect to Z, x must be in some Σ. (The point is that for large enough h ${}_r\tilde{j}^\infty(f)^{-1}(S(A^h(Z),P,N))$ is empty.) Then (194) implies that $T(t_1, \ldots t_a, x) \in Q_1 \times \ldots \times Q_a$, a contradiction. This proves $(189)_{\alpha_0}$. QED

§III.C. $\underline{(189)_{\alpha+1}} \Rightarrow \underline{(189)_\alpha}$ $\underline{(0\le\alpha<\alpha_0)}$.

For all of §III.C. α will be a fixed integer satisfying $0\le\alpha<\alpha_0$. Assume $(189)_{\alpha+1}$. To prove $(189)_\alpha$, choose an isotopy class of fibered concordances satisfying (187) and let $F = (h,f,p_3)$ be a representative for it which satisfies the conditions of $(189)_{\alpha+1}$ for some open cover $\{\mathcal{O}_j\}_{j=1}^{a}$ of D^s.

§III.C.1. Reduction to Claim 113.

Let r and m be the rank and level of Z_α. For each admissible pair of orderings (D,R) for Z_α, set

$$K^{(D,R)}(F) = \bigcup_{j=1}^{a}\bigcup_{i=1}^{r}\left\{ x \in W_\alpha^{(D,R)}(F) \;\middle|\; p_3\pi_i x\in\overline{\mathcal{O}}_j \text{ and for some } t\in I\ f(t,p_{2,3}\pi_i x) \in Q_j \right\}.$$

If each of the sets $K^{(D,R)}(F)$ were empty then F would already satisfy the conditions of $(189)_\alpha$ (for the cover $\{\mathcal{O}_j\}$). The plan will be to modify F by a fibered isotopy so as to make them empty, one by one. That is, for a fixed pair (D,R) we will show:

CLAIM 113. (Assuming F and $\{O_j\}$ satisfy the conditions of $(189)_{\alpha+1}$ as well as (187)) there is another representative F' satisfying the additional condition $K^{(D,R)}(F') = \emptyset$, and such that for any pair (D',R') $K^{(D',R')}(F) = \emptyset \Rightarrow K^{(D',R')}(F') = \emptyset$.

The proof of Claim 113 will occupy the rest of §III.C.

§III.C.2. Beginning of Proof of Claim 113.

The rest of §III.C corresponds to the discussion at the end of §D.5 of the Introduction, the correspondence being given by the following "dictionary":

$$p_{2,3}\pi_{i_0} W_\alpha^{(D,R)}(F) \longleftrightarrow \Sigma$$

$$\pi_{i_0} W_\alpha^{(D,R)}(F) \longleftrightarrow \tilde{\Sigma}$$

$$\pi_{i_0} K^{(D,R)}(F) \longleftrightarrow K.$$

The first step is to prove:

CLAIM 114. $K^{(D,R)}(F)$ is compact.

PROOF of Claim: $K^{(D,R)}(F)$ is a closed (and open) subset of the disjoint union

$$\bigcup_{(D',R')} K^{(D',R')}(F) .$$

This union is the image, under $(x,t) \longmapsto x$, of the set

$$\bigcup_{j=1}^{a} \bigcup_{i \in I(Z_\alpha)} K_{j,i}(F) ,$$

where

$$K_{j,i}(F) \underset{\text{def}}{=} \left\{ (x,t) \in W_\alpha(F) \times I \;\middle|\; p_3 \pi_i x \in \bar{\mathcal{O}}_j \text{ and } f(t, p_{2,3}\pi_i x) \in Q_j \right\} .$$

Therefore it suffices to show that each $K_{j,i}(F)$ is compact, i.e. that $K_{j,i}(F)$ is closed in $((I \times P)^r \times D^s) \times I$.

Let (x,t) be a limit point of $K_{j,i}(F)$. By continuity

$$p_3 \pi_i x \in \bar{\mathcal{O}}_j \tag{196}$$

$$f(t, p_{2,3}\pi_i x) \in Q_j . \tag{197}$$

We must show that $x \in W_\alpha(F)$. Lemma 90 implies that either

(198) $$x \in {}_r\tilde{j}^\infty(f)^{-1}S(Z_\alpha,P,N)$$

or

(199) $$x = \phi^*y,\ y \in {}_{r'}\tilde{j}^\infty(f)^{-1}S(D_\phi(Z_\alpha),P,N)$$

for some r' and some surjection $\phi\colon \{1,\ldots r\} \longrightarrow \{1,\ldots r'\}$. If (199) holds, let $D_\phi(Z_\alpha) = Z_\beta$ (where $\beta>\alpha$ necessarily, by Lemma 77 (x) and (172)). We have

$$F(t,p_{2,3}\pi_{\phi(i)}y) = F(t,p_{2,3}\pi_i x) \in I\times Q_j\times D^s ,$$

so $\pi_{\phi(i)}y \in I\times \mathrm{int}\, P\times D^s$ and by Lemma 92 $y \in S_\beta(F)$. Since $\beta>\alpha$ this contradicts the fact that F and $\{\mathcal{O}_j\}$ satisfy the conditions of $(189)_{\alpha+1}$. (We have $f(t,p_{2,3}\pi_{\phi(i)}y) \in Q_j$ and $p_{2,3}\pi_{\phi(i)}y \in (P\times\overline{\mathcal{O}}_j) \cap p_{2,3}\pi_{\phi(i)}S_\beta(F)$.)

Therefore (198) must hold. Use Lemma 92 again: the fact that $\pi_i x \in I\times \mathrm{int}\, P\times D^s$, which follows from (197), implies that $x \in S_\alpha(F)$. If $x \notin W_\alpha(F)$ then by Lemma 101 there exist $\beta>\alpha$ and a morphism $\phi\colon Z_\alpha \longrightarrow Z_\beta$ such that $x \in \phi^*S_\beta(F)$. But this again contradicts the fact that F and $\{\mathcal{O}_j\}$ satisfy the conditions of $(189)_{\alpha+1}$: If $x = \phi^*y$, $y \in S_\beta(F)$, then

$$f(t,p_{2,3}\pi_{\phi(i)}y) \in Q_j \quad \text{and}$$

$$p_{2,3}\pi_{\phi(i)}y \in P\times\overline{O}_j \ .$$

Therefore $x \in W_\alpha(F)$. This proves Claim 114. QED.

Choose an element $i_0 \in I(Z_\alpha) = \{1,\dots r\}$ which is maximal with respect to D in its $\sim$-class and maximal with respect to R in its $\simeq$-class. (Existence of such an i_0 follows easily from the fact that Z_α is good. Z_α is good by (190) and Lemma 85.)

Next it will be necessary to adjust F slightly so that (in the language of §Intro. D.5) the set $K \subset \tilde{\Sigma}$ does not touch $0\times P\times D^s$. This is certainly necessary - otherwise it will be impossible to find a function $\phi\colon P\times D^s \longrightarrow (0,1]$ whose graph cuts below K.

CLAIM 115. Without loss of generality (in proving Claim 113) we can assume

(200) $$p_1\pi_{i_0}(x) > 0 \quad \text{for all} \quad x \in K^{(D,R)}(F) \ .$$

PROOF of Claim: First of all, (200) automatically holds if the rank r of Z_α is greater than one. For then by Lemma 69 (condition (149)) there exists $i \in I(Z_\alpha) = \{1,\cdots r\}$ such that $i \neq i_0$ and either $i \sim_{Z_\alpha} i_0$ or $i \simeq_{Z_\alpha} i_0$. Thus by the way in which i_0 was chosen either iDi_0 or iRi_0. By Definition 106 either $\pi_i x$ is below $\pi_{i_0} x$ or $F\pi_i x$ is below $F\pi_{i_0} x$, for any $x \in W_\alpha^{(D,R)}(F)$. In either case $\pi_{i_0} x \notin 0\times P\times D^s$, that is, $p_1\pi_{i_0} > 0$.

So assume $r=1$. Recall the sets $K_{j,i}(F)$ in the proof of Claim 114. In this case i is always one. For each $j(1\le j\le a)$ the function

$$K_{j,1}(F) \longrightarrow I$$

$$(x,t) \longmapsto t$$

is a continuous function on a compact set, and positive. Choose $\epsilon>0$ such that

$$(201) \qquad (x,t) \in K_{j,1}(F) \Rightarrow t>\epsilon$$

for all j.

The idea behind the proof of Claim 115 will be to find a function $\phi : P\times D^s \longrightarrow (0,1]$ which (besides satisfying "sunniness" conditions) separates the set

$$A = K^{(D,R)}(F) \cap (0\times P\times D^s)$$

(which we are trying to eliminate) from the set

$$B = \left\{ \text{points in } \bigcup_{j=1}^{a} K_{j,1}(F) \text{ which are above } A \right\}$$

Then the associated fibered isotopy F^u will make B disappear:

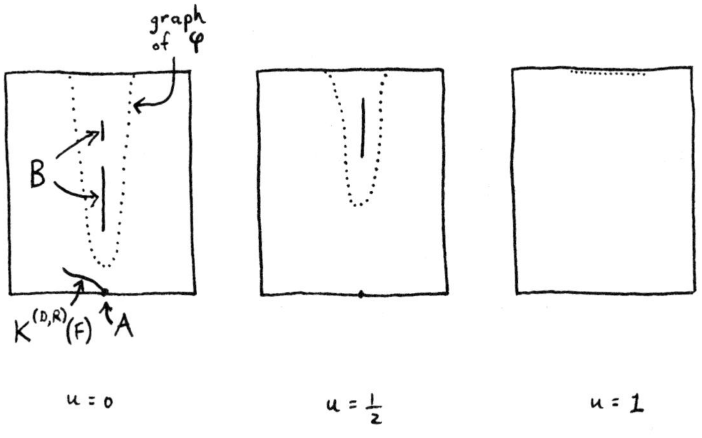

Since B is empty for F^1, so will A be, by definition of $K^{(D,R)}(F^1)$.

More formally, we want a smooth function

$$P\times D^s \xrightarrow{\ \phi\ } (0,1],$$

satisfying not only (49), (50), and (51) (§1.C) but also

$$(202)\qquad \begin{cases} \phi(p_{2,3}x) = \epsilon & \text{for all} \\ x \in K^{(D,R)}(F) & \text{such that } p_1x = 0 \ . \end{cases}$$

Assume that we have such a ϕ. We claim that ϕ yields a fibered isotopy F^u with $F^\circ = F$, such that (187) is preserved (that is, holds for all u), and F^1 still satisfies the conditions of $(189)_{\alpha+1}$, and F^1 satisfies (200). (Note that $K^{(D',R')}(F) = \emptyset \Rightarrow K^{(D',R')}(F^1) = \emptyset$, since Z_α has rank one.) To see this, define ϕ^u as in Lemma 30 and use Lemma 29 to get an isotopy F^u. Write $F^u = (h^u, f^u, p_3)$.

First of all, each F^u satisfies (187), by (48), because $F^\circ = F$ does.

Next check the conditions of $(189)_{\alpha+1}$ for F^1. Because F was in general position with respect to Z, the same is true of F^1 by Lemma 55 ((118)). The singular sets $S_\beta(F')$ are also described by Lemma 55 (in terms of the $S_\beta(F)$). Namely for any Z_β, say of rank r', we have

(203) $$((t_1,x_1),\cdots(t_{r'},x_{r'}),y) \in S_\beta(F') \iff$$

$$((t_1\phi(x_1,y),x_1),\cdots(t_{r'}\phi(x_{r'},y),x_{r'}),y) \in S_\beta(F) .$$

Now, to finish checking the conditions of $(189)_{\alpha+1}$ for F^1 and $\{\mathcal{O}_j\}$ we must show that

$$f^1(I\times(P\times\overline{\mathcal{O}}_j \cap p_{2,3}\pi_i S_\beta(F^1))) \cap Q_j = \emptyset$$

for all $\beta>\alpha$, $i \in I(Z_\beta)$, and $1\le j\le a$. Suppose this fails for some β, i, and j. Then there exist $t \in I$ and

$$x = ((t_1,x_1),\cdots(t_{r'},x_{r'}),y) \in S_\beta(F')$$

such that

$$y \in \overline{\mathcal{O}}_j$$

$$f(t,x_i,y) \in Q_j .$$

But then by (46) and (203) the point

$$x' = ((t_1\phi(x_1,y),x_1),\cdots(t_{r'}\ \phi(x_{r'},y),x_{r'}),y)$$

is in $S_\beta(F)$ and satisfies

$$y \in \overline{O}_j$$

$$f(t\phi(x_i,y),x_i,y) \in Q_j \ .$$

This contradicts the hypothesis for F.

To check (200) for F^1 suppose $x \in K^{(D,R)}(F^1)$ and $p_1x = 0$. Set $x = (0,\ x_1,y) \in I\times P\times D^s$, and say $(x,t) \in K_{j,1}(F^1)$. Then it is easy to see that if we define x' as above then

$$x' \in K^{(D,R)}(F)$$

$$(x',t\phi(x_1,y)) \in K_{j,1}^{(D,R)}(F) \ .$$

(Of course in this case $x' = x$.) Thus

$$t\phi(x_1,y) > \epsilon \quad \text{by (201)}$$

$$= \phi(x_1,y) \quad \text{by (202)} \ ,$$

a contradiction.

To construct ϕ, choose a very small neighborhood N of the compact set

$$A \underset{\text{def}}{=} p_{2,3} \left\{ x \in K^{(D,R)}(F) \;\middle|\; p_1 x = 0 \right\}$$

in $P\times D^s$. (We will soon say how small.) Choose a smooth function

$$P\times D^s \xrightarrow{\psi} I$$

supported in N, equal to 1 on A. Set

$$\phi = 1 - (1 - \epsilon)\,\psi \,.$$

ϕ will satisfy (202), obviously.

ϕ will satisfy (49), provided $N \subset \text{int}\, P\times D^s$.

For (50) and (51) it will suffice if

(204) $$Y_0(F) \cap ([\epsilon,1]\times N) = \emptyset \quad .$$

For if (t,x,y) satisfies the hypothesis of (50) then $(t,x,y) \in \pi_1\, S_0(F)$, so that by (204) $(t,x,y) \notin [\varepsilon,1]\times N$, so that either

$$t < \epsilon \le \phi(x,y)$$

or

$$(x,y) \notin N$$

$$\phi(x,y) = 1 ,$$

contradicting the hypothesis of (50). Likewise if (t_0,x_0,y_0) satisfies the hypothesis of (51) then $(t_0,x_0,y_0) \in \pi_1 S_1(F)$ so that by (204) $(t_0,x_0,y_0) \notin [\epsilon,1]\times N$. Again either

$$t_0 < \epsilon \leq \phi(x_0,y_0)$$

or

$$\phi(x_0,y_0) = 1 ,$$

contradicting the hypothesis of (51) either way.

We must find a neighborhood N of A satisfying (204). As was shown in the proof of Lemma 107, $Y_0(F)$ is closed in $I\times \text{int } P\times D^s$. Therefore it suffices to check that

$$Y_0(F) \cap ([\epsilon,1]\times A) = \emptyset .$$

Suppose this is not the case, so that there exist

$$\alpha' \geq 0 \ , \ i' \in I(Z_{\alpha'}), \ x' \in S_{\alpha'}(F), \ x \in K^{(D,R)}(F)$$

such that

$$p_{2,3}\pi_{i'}x' = p_{2,3}\pi_1 x$$

$$p_1\pi_{i'}x' \geq \epsilon$$

$$p_1\pi_1 x = 0 \ ,$$

Lemma 98 (with $i=1$) yields as usual

$$Z_\alpha \xrightarrow[\cong]{\phi} Z_{\alpha''} \xleftarrow{\phi'} Z_{\alpha'}$$

(ϕ is an isomorphism because $x \in K^{(D,R)}(F) \subset W_\alpha(F)$). Then

$$\epsilon \leq p_1\pi_{i'}x' = p_1\pi_{\phi'(i')}x'' = p_1\pi_{\phi^{-1}\phi'(i')}x = p_1\pi_1 x = 0 \quad ,$$

since $\phi^{-1}\phi'(i') \in I(Z_\alpha) = \{1\}$. This contradiction completes the proof that ϕ exists, and so the proof of Claim 115.

QED

For the rest of §III.C we assume (200).

We will obtain Claim 113 from:

CLAIM 116: (Assuming the hypotheses of Claim 113, and (200), too) there is a smooth fibered sunny collapse

$$P\times D^s \xrightarrow{\phi^u} (0,1] \quad (0\le u\le 1)$$

satisfying

$$(205)\qquad \begin{cases} \text{for all } x \in K^{(D,R)}(F) \\ \phi^1(p_{2,3}\pi_{i_0}x) < p_1\pi_{i_0}x \ . \end{cases}$$

(This is the main sunny collapsing step in the proof — the one which was described in §D.5 of the introduction.)

Before proving Claim 116 we show that it implies Claim 113. Like the proof of Claim 115 this is a straightforward application of Lemmas 29 and 55. Let ϕ^u be as in Claim 116 and let F^u be the associated isotopy (Lemma 29). Write $F^u = (h^u,f^u,p_3)$. Write $F^u = (h^u,f^u,p_3)$.

One can check that F' satisfies the conditions of $(189)_{\alpha+1}$ just as in the proof of Claim 115.

To check that $K^{(D,R)}(F^1) = \emptyset$, suppose that x is in $K^{(D,R)}(F^1)$. Then x' (defined as on page 272 with $r' = r$) is easily seen to be in $K^{(D,R)}(F)$. Thus

$$\phi(p_{2,3}\pi_{i_0}x') < p_1\pi_{i_0}x' \quad \text{by (205)}$$

$$= \phi(p_{2,3}\pi_{i_0}x')p_1\pi_{i_0}x$$

$$\leq \phi(p_{2,3}\pi_{i_0}x') ,$$

a contradiction.

For the last statement of Claim 113, just observe that if $x \in K^{(D',R')}(F^1)$ then x' (defined as on page 272, with $r' = r$) is in $K^{(D',R')}(F)$.

§III.C.3. Reduction to $(206)_0$.

It remains to prove Claim 116.

PROOF of Claim 116: The proof is inductive. Roughly speaking, we construct ϕ^u in a neighborhood of the set $p_{2,3}Y_\beta$, first for $\beta=\alpha$ and then inductively for smaller and smaller β, ending with $\beta=0$. More precisely we will use as inductive hypothesis, for $\alpha\geq\beta\geq 0$:

$(206)_\beta$ There is a smooth homotopy

$$P\times D^s \xrightarrow{\phi^u_\beta} (0,1], \qquad 0\leq u\leq 1$$

such that

(i) ϕ^u_β satisfies (205)

(ii) ϕ^u_β satisfies (34)

(iii) ϕ^u_β satisfies (40) in the special case: $(x',y) \in p_{2,3}Y_\beta$

(iv) ϕ^u_β satisfies (41) in the special case: $(x,y) \in p_{2,3}Y_\beta$

(v) For each $(x,y) \in P\times D^s$ either $\phi^u_\beta(x,y) = 1$ for all u or $\frac{\partial}{\partial u}\phi^u_\beta(x,y) < 0$ for all u.

Thus (i) means that the graph of ϕ^u_β cuts below $K^{(D,R)}(F)$, and (iii) and (iv) mean that, where no higher stratum $p_{2,3}\pi_i W^{(D',R')}_\gamma(F)$, $\gamma = \beta-1,\ldots 0$, is involved, ϕ^u_β is "sunny". Then $\phi^u_{\beta-1}$ represents an improvement on ϕ^u_β ; it satisfies sunniness a little more generally. The improvement (III.C.5 below) is made by leaving ϕ^u_β unchanged in a neighborhood of $p_{2,3}Y_\beta$ and, roughly, making it take the value one outside a slightly larger neighborhood. Unfortunately, however, it is not just a matter of making that larger neighborhood small enough; we have to avoid interference between different strata $p_{2,3}\pi_i W^{(D',R')}_\gamma(F)$ and $p_{2,3}\pi_j W^{(D',R')}_\gamma(F)$ (with, say, $iR'j$). This leads to an induction, in §III.C.5.b, with respect to the ordering R' .

The proof of Claim 116 will be complete when $(206)_0$ is established: Then ϕ^u_0 satisfies (40) and (41) unconditionally and is the desired sunny collapse ϕ^u . Indeed, if (40) is violated, then there exists (u,t,x,x',y) such that

$$0 \le t \le \phi^u_0(x,y)$$

$$F(\phi^u_0(x',y),x',y) \text{ is below } F(t,x,y) .$$

Thus

$$((t,x),(\phi_0^u(x',y),x'),y) \in {}_2\tilde{j}^\infty(f)^{-1}S(Z_0,P,N) .$$

If the point is in $S_0(F)$ then $(x',y) \in p_{2,3}Y_0$, contradicting $(206)_0$ (iii). Thus $x' \in \partial P$, so that by (34) applied to ϕ_0^u we have

$$\phi_0^u(x',y) = 1 .$$

Therefore

$$F(\phi_0^u(x',y),x',y) \in 1\times N\times D^s$$

cannot be below any point in $I\times N\times D^s$, a contradiction.

Likewise if (t,x,y,v) violates (41) (for ϕ_0^u) then the fact that $DF\cdot v = \frac{\partial}{\partial t}$ implies

$$(t,x,y) \in {}_1\tilde{j}^\infty(f)^{-1}S(Z_1,P,N), \quad \text{and}$$

$(206)_0$ (iv) rules out the possibility that $(t,x,y) \in S_1(F)$. Thus $x \in \partial P$. But then $DF\cdot v = \frac{\partial}{\partial t} \Longrightarrow v = \frac{\partial}{\partial t}$, so that

$$d(\phi_0^u \circ p_{2,3} - p_1)\cdot v = 0-1<0$$

and (41) is not violated after all.

§III.C.4. Proof of $(206)_\alpha$.

We now begin the induction with respect to β by proving $(206)_\alpha$. The function ϕ^u_α will have the form

$$1 - u(1-\phi) ,$$

so that instead of verifying that ϕ^u_α satisfies (34) and in special cases (40) and (41), we have only to verify that ϕ satisfies (49) and in corresponding special cases (50) and (51). The last condition $(206)_\alpha$ (v) is automatic because of the special form of ϕ^u_α .

Geometrically the key idea in constructing ϕ^u_α is that, by the way in which i_0 was chosen, $\pi_{i_0} W_\alpha^{(D,R)}(F)$ has no points of Y_0 above it in $I\times P\times D^s$ and $F(\pi_{i_0} W_\alpha^{(D,R)}(F))$ has no points of $F(I\times P\times D^s)$ above it in $I\times N\times D^s$. (In the language of §Intro. D.5, $\hat{\tilde{\Sigma}} \cup \tilde{\Sigma}$ is disjoint from $\pi_1 S_0$.)

<u>Case 1</u>. $i_0 \notin \Delta_{Z_\alpha}$. ($\Delta_{Z_\alpha}$ is defined in Def. 68, §II.C.1). Consider the sets

$$X_i^{(D',R')} \underset{\text{def}}{=} p_{2,3}\pi_i W_\alpha^{(D',R')}(F) \subset P\times D^s$$

where $i \in I(Z_\alpha)$ and (D',R') is an admissible pair for Z_α. According to Lemmas 99 and 110, $X_i^{(D',R')}$ is a homeomorphic image of $W_\alpha{}^{(D',R')}$ and is a closed subset of the open set

$$(\text{int}P)\times D^s - p_{2,3}Y_{\alpha+1} \subset P\times D^s ,$$

and $X_i^{(D',R')}$ is disjoint from $X_{i_0}^{(D,R)}$ except when it is equal to it. Set

$$C_0 = \partial P\times D^s \cup p_{2,3}Y_{\alpha+1} \cup \bigcup_{X_i^{(D',R')} \neq X_{i_0}^{(D,R)}} X_i^{(D',R')} ,$$

a closed subset of $P\times D^s$. Choose a smooth function

$$\phi : P\times D^s \longrightarrow (0,1]$$

such that

(207) $$\phi \equiv 1 \text{ near } C_0$$

(208) $$\phi(p_{2,3}\pi_{i_0}x) > p_1\pi_i x \quad \text{for all} \quad x \in W^{(D,R)}(F) \quad \text{and} \quad iDi_0 \quad ,$$

(209) $$\phi(p_{2,3}\pi_{i_0}x) < p_1\pi_{i_0}x \quad \text{for all} \quad x \in K^{(D,R)}(F) \quad .$$

To find such a function on $P\times D^s$ it suffices to find one locally near each point and then use a partition of unity. Locally there is always a constant function satisfying (207)-(209). In fact, outside of the compact set $p_{2,3}\pi_{i_0}K^{(D,R)}(F)$ the constant 1 will do, since for iDi_0 and $x \in W_\alpha^{(D,R)}(F)$ we have

(210) $$p_1\pi_i x < p_1\pi_{i_0}x \le 1 \; .$$

Near a point $p_{2,3}\pi_{i_0}x_0$ in that compact set the constant has only to be less than $p_1\pi_{i_0}x_0$, positive, and greater than each $p_1\pi_i x_0$ for iDi_0, since the functions

$$p_{2,3}\pi_{i_0}x \longmapsto p_1\pi_i x$$

on $X_{i_0}^{(D,R)}$ are continuous. Such a constant exists by (200) and (210).

Now we will verify that ϕ_α^u satisfies the conditions of $(206)_\alpha$.

$\phi_\alpha^1=\phi$ satisfies $(206)_\alpha$(i) by (209).

ϕ_α^u satisfies $(206)_\alpha$(ii) by (207) .

For (iii), assume

(211) $F(t,x,y)$ is below $F(t',x',y)$, and

(212) $$(x,y) \in p_{2,3}Y_\alpha$$

Now, (212) implies that $(x,y) \in C_0 \cup X_{i_0}^{(D,R)}$. Also, (211) implies $F(t,x,y) \notin 1\times N\times D^s$, hence $t<1$.

If $(x,y) \in C_0$, then by (207) we have

$$\phi(x,y) = 1>t ,$$

so that (50) is not violated.

Now suppose $(x,y) \in X_{i_0}^{(D,R)}$; say

$$(x,y) = p_{2,3}\pi_{i_0}\tilde{x} , \tilde{x} \in W_\alpha^{(D,R)}(F) .$$

Statement (211) implies that the point

$$\tilde{x}' \underset{\text{def}}{=} ((t,x),(t',x'),y)$$

is in $S_0(F)$. Apply Lemma 98 with 0 for α', $\tilde{x}$ for x, $\tilde{x}'$ for x', i_0 for i, 1 for i'. The result is a set $Z_{\alpha''}$ with morphisms

$$Z_\alpha \xrightarrow{\phi} Z_{\alpha''} \xleftarrow{\phi'} Z_0$$

and a point $\tilde{x}'' \in S_{\alpha''}(F)$ such that

$$\tilde{x} = \phi^*\tilde{x}'' \;,\; \tilde{x}' = \phi'^*\tilde{x}'' \;.$$

Since $\tilde{x} \in W_\alpha(F)$, ϕ is an isomorphism. Set

$$i = \phi^{-1}\phi'(1) \;,\; j = \phi^{-1}\phi'(2) \;.$$

Then by Lemma 100 $i \sim_{Z_\alpha} i_0$, so that by the way in which i_0 was chosen either iDi_0 or $i=i_0$.

If iDi_0, then by (208)

$$\begin{aligned} \phi(x,y) &= \phi(p_{2,3}\pi_{i_0}\tilde{x}) \\ &> p_1\pi_i\tilde{x} = t \;, \end{aligned}$$

so that (50) is not violated.

If $i=i_0$ then $F(\pi_{i_0}\tilde{x}) = F(\pi_i\tilde{x}) = F(\pi_1\tilde{x}') = F(t,x,y)$ is below $F(t',x',y) = F(\pi_2\tilde{x}') = F(\pi_j\tilde{x})$ so that by Lemma 100 $i_0 R j$, in contradiction to the way in which i_0 was chosen.

To verify $(206)_\alpha(\text{iv})$ we must check (51) assuming:

$$(x_0,y_0) \in p_{2,3}Y_\alpha \ .$$

As above, this implies $(x_0,y_0) \in C_0 \cup X_{i_0}^{(D,R)}$. Let the tangent vector v be as in the hypothesis of (51). If $(x_0,y_0) \in C_0$ then by (207) the hypothesis of (51) is impossible. So as above we can assume that $(x_0,y_0) \in X_{i_0}^{(D,R)}$. Write

$$(x_0,y_0) = p_{2,3}\pi_{i_0}\tilde{x} \ , \ \tilde{x} \in W_\alpha^{(D,R)}(F) \ .$$

The hypothesis concerning v in (51) implies that the point

$$\tilde{x}' \underset{\text{def}}{=} (t_0,x_0,y_0)$$

is in $S_1(F)$. Apply Lemma 98 with $\tilde{x}$ for x, $\tilde{x}'$ for x'. 1 for α', i_0 for i, and 1 for i'. As usual this yields $\alpha'' \geq 0$, morphisms

$$Z_\alpha \xrightarrow{\ \phi\ } Z_{\alpha''} \xleftarrow{\ \phi'\ } Z_1 \ ,$$

and a point $\tilde{x}'' \in S_{\alpha''}(F)$ such that

$$\tilde{x} = \phi^*\tilde{x}'' , \tilde{x}' = \phi'^*\tilde{x}'' .$$

Since $\tilde{x} \in W_\alpha(F)$, ϕ is an isomorphism. Set

$$i = \phi^{-1}\phi'(1).$$

$i \sim_{Z_\alpha} i_0$ by Lemma 100, because

$$p_{2,3}\pi_i\tilde{x} = (x_0,y_0) = p_{2,3}\pi_{i_0}\tilde{x} .$$

Thus either iDi_0 or $i=i_0$. If iDi_0, then by (208)

$$\phi(x_0,y_0) = \phi(p_{2,3}\pi_{i_0}\tilde{x})$$

$$> p_1\pi_i\tilde{x} = t ,$$

so that (51) is not violated. If $i=i_0$ then the existence of v implies $i_0 \in \Delta_{Z_\alpha}$, contrary to assumption.

Case 2. $i_0 \in \Delta_{Z_\alpha}$. In this case we must construct ϕ a little more carefully. Let ξ and η be the vector fields ξ_{α,i_0} and η_{α,i_0} defined in §II.D.2.b.iv (discussion preceding Lemma 104). We will impose the extra condition

$$(213)\quad \begin{cases} (d\phi)\cdot\eta_x < \dfrac{1-\phi p_{2,3}\pi_{i_0}x}{1-p_1\pi_{i_0}x}(dp_1)\cdot\xi_x \ , \\ \text{for all} \quad x \in W_\alpha^{(D,R)}(F) \quad \text{such that} \\ \phi\; p_{2,3}\pi_{i_0}x < p_1\pi_{i_0}x < 1 \end{cases}$$

on ϕ, along with (207), (208), and (209).

Assuming for the moment that ϕ can be chosen so as to satisfy (207), (208), (209), and (213), let us finish verifying the conditions of $(206)_\alpha$. As in Case 1 we end up having only to verify (51) in the situation:

$$\tilde{x} \in W_\alpha^{(D,R)}(F)$$

$$(t_0,x_0,y_0) = \pi_{i_0}\tilde{x} \ .$$

The hypothesis of (51) implies that

$$v = \xi_{\tilde{x}} \quad \text{and}$$

$$\phi(p_{2,3}\pi_{i_0}\tilde{x}) \le p_1\pi_{i_0}\tilde{x} < 1 .$$

Thus (213) implies the conclusion of (51).

As in Case 1, ϕ can be constructed by using a partition of unity. Again, in the complement of $p_{2,3}\pi_{i_0}K^{(D,R)}(F)$ the constant function 1 satisfies (207), (208), (209), and (213). ((213) is vacuous wherever $\phi = 1$.) Given $x_0 \in K^{(D,R)}(F)$, choose a function ϕ near $p_{2,3}\pi_{i_0}x_0$ whose value at that point is less than $p_1\pi_{i_0}x_0$, positive, and greater than $p_1\pi_i x_0$ for each iDi_0, and whose derivative with respect to η_{x_0} is less than

$$\frac{1-\phi p_{2,3}\pi_{i_0}x_0}{1-p_1\pi_{i_0}x_0}(dp_1)\cdot\xi_{x_0} .$$

Then as in Case 1 if ϕ is restricted to a sufficiently small open set near $p_{2,3}\pi_{i_0}x_0$ it will satisfy (208) and (209) (and vacuously (207)). Moreover, it will satisfy (213).

Now we have local solutions of the problem defined on a finite open cover of $P\times D^s$. Patch them together to get a global solution by using a partition of unity. In order that the result should satisfy (213) it is enough that each of the functions μ in the partition of unity should satisfy

$$(d\mu)\cdot\eta_x = 0$$

for all $x \in W_\alpha^{(D,R)}(F)$. This condition is easy to arrange using Lemma 104.

Now the proof of $(206)_\alpha$ is complete (assuming that F and $\{O_j\}$ satisfy the conditions of $(189)_{\alpha+1}$). To finish the proof of $(189)_\alpha$ we have to prove that $(206)_{\beta+1} \Rightarrow (206)_\beta$ for $0\le\beta<\alpha$.

§III.C.5. $\underline{(206)_{\beta+1} \Rightarrow (206)_\beta}$.

Fix $\beta(0\le\beta<\alpha)$ for the rest of §III.C. Assume that

$$P\times D^s \xrightarrow{\phi^u_{\beta+1}} (0,1] , \quad 0\le u\le 1$$

satisfies the requirements of $(206)_{\beta+1}$. We are going to construct

$$P\times D^s \xrightarrow{\phi^u_\beta} (0,1] , \quad 0\le u\le 1$$

so as to satisfy the requirements of $(206)_\beta$.

Set

$$C = \partial P \times D^s \cup p_{2,3} Y_{\beta+1}$$

a closed set in $P \times D^s$. By assumption ($(206)_{\beta+1}$) $\phi^u_{\beta+1}$ satisfies (40) and (41) for $(x',y) \in C$ and $(x,y) \in C$ respectively, and consequently also for points in a neighborhood N of C. (The set

$$\left\{ (t,x,t',x',y) \in (I \times P)^2 \times D^s \;\middle|\; \begin{array}{l} \text{either} \\ (40) \text{ fails for } \phi^u_{\beta+1} \text{ at } (t,x,t',x',y) \text{ or} \\ (41) \text{ fails for } \phi^u_{\beta+1} \text{ at } (t,x,y) = (t',x',y) \end{array} \right\}$$

is compact.)

Recall that $W_\beta(F)$ is a disjoint union

$$W_\beta(F) = \bigcup_{(D',R')} W_\beta^{(D',R')}(F) ,$$

where (D',R') ranges over the admissible pairs for Z_β, and that the images

$$X_i^{(D',R')} \underset{\text{def}}{=} p_{2,3} \pi_i W_\beta^{(D',R')}(F)$$

are closed in $P \times D^s - C$ and are either disjoint or equal (Lemma 99).

We fix some notation: Say that two of these pairs (D',R') and (D'',R'') are <u>equivalent</u> if there is an automorphism of Z_β taking (D',R') to (D'',R''). Choose representatives for the equivalence classes:

$$(D_1,R_1),\cdots(D_c,R_c)$$

($c \geq 1$ being some integer). For each $k(1 \leq k \leq c)$ number the $\sim_{Z_\beta}$-classes

$$C_{k,1},\cdots C_{k,b}$$

($b \geq 1$ some integer) in such a way that

$$(214) \qquad \left\{ \begin{array}{l} i_1 \in C_{k,j_1},\ i_2 \in C_{k,j_2},\ i_1 R_k i_2 \\ \\ \qquad \Longrightarrow j_2 < j_1 . \end{array} \right.$$

(This is possible because Z_β is good — Exercise.) Now the images

$$X_{k,j} \overset{=}{\text{def}} X_i^{(D_k,R_k)}, \text{ any } i \in C_{k,j}$$

$$(1 \leq k \leq c\ ,\ 1 \leq j \leq b)$$

are disjoint and include all of the $X_i^{(D',R')}$ (Lemma 99).

§III.C.5.a. Reduction to Claim 117.

To construct ϕ_β^u we will use:

CLAIM 117. There exist smooth homotopies

$$P\times D^s \xrightarrow{\mu_{k,j}^u} I \qquad 0\le u\le 1$$

for $1\le k\le c$, $1\le j\le b$, such that, setting

$$(215)\qquad \phi_{k,j}^u = 1 - \mu_{k,j}^u\cdot(1-\phi_{\beta+1}^u),$$

we have:

$$(216)\qquad \mu_{k,j}^u \equiv 1 \text{ in a neighborhood of } C$$

$$(217)\qquad \mathrm{supp}(\mu_{k,j}^u) \subset N$$

(218) if $\tilde{x} \in W_\beta^{(D_k,R_k)}(F)$, $i_1 \in C_{k,j_1}$, $i_2 \in C_{k,j_2}$, and $j_2 < j_1$, then

$$\mu^{u}_{k,j_1} p_{2,3} \pi_{i_1} \tilde{x} \neq 0 \implies$$

$$\mu^{\hat{u}}_{k,j_2} p_{2,3} \pi_{i_2} \tilde{x} = 1 \quad \text{for all} \quad \hat{u} \quad (0 \le \hat{u} \le 1) .$$

(219) if $\tilde{x} \in W_{\beta}^{(D_k, R_k)}(F)$, $i \in C_{k,j} \cap \Delta_{Z_\beta}$,

and $\Phi^{u}_{k,j}(p_{2,3}\pi_i \tilde{x}) = p_1 \pi_i \tilde{x}$, then

$$(d\Phi^{u}_{k,j}) \cdot \eta_{\beta,i} < (dp_1) \cdot \xi_{\beta,i}$$

[where $\eta_{\beta,i}$ and $\xi_{\beta,i}$ are as defined in §II.D.2.b.iv]

(220) for each $(x,y) \in P \times D^s$ either

$\Phi^{u}_{k,j}(x,y) = 1$ for all u or

$\frac{\partial}{\partial u} \Phi^{u}_{k,j}(x,y) < 0$ for all u .

Before proving Claim 117, we show that it implies the existence of ϕ^{u}_{β} satisfying the conditions of $(206)_\beta$.

By (216) and (215) $\phi^u_{k,j}$ coincides with $\phi^u_{\beta+1}$ on a neighborhood of C. Since the $X_{k,j}$ are disjoint closed subsets of $P\times D^s - C$, it is possible (using a partition of unity in $P\times D^s$) to construct ϕ^u_β such that

(221) $$\phi^u_\beta \equiv \phi^u_{\beta+1} \text{ in a neighborhood of } C$$

(222) $$\phi^u_\beta \equiv \phi^u_{k,j} \text{ in a neighborhood of } X_{k,j} .$$

$(206)_\beta$(v) then follows from $(206)_{\beta+1}$(v) and (220). For the rest of $(206)_\beta$:

$(206)_\beta$(i) follows from $(206)_{\beta+1}$(i) and (221).

$(206)_\beta$(ii) follows from $(206)_{\beta+1}$(ii) and (221).

For $(206)_\beta$(iii) suppose $(x',y) \in p_{2,3}Y_\beta$. If $(x',y) \in p_{2,3}Y_{\beta+1}$ then (40) follows from $(206)_{\beta+1}$(iii) and (221). Otherwise we must have $(x',y) \in X_{k,j}$ for some k and j . So assume:

(223) $$(x',y) \in X_{k,j}$$

(224) $$\begin{cases} F(t',x',y) \text{ is below } F(t,x,y) \\ \phi^u_\beta(x',y) = t' \\ 0 \le t \le \phi^u_\beta(x,y) \end{cases} .$$

Statement (223) means that there exist $\tilde{x} \in W_{\alpha}^{(D_k,R_k)}(F)$ and $i \in C_{k,j}$ such that

$$(x',y) = p_{2,3}\pi_i\tilde{x} .$$

From (224) it follows that $(t',x',t,x,y) \in S_0(F)$. Then using Lemma 98 we see that there exist i_1 and i_2 in $I(Z_\beta)$ such that

$$(t',x',y) = \pi_{i_1}\tilde{x}$$

$$(t,x,y) = \pi_{i_2}\tilde{x} .$$

By definition of $W_{\beta}^{(D_k,R_k)}(F)$ we have

(225) $$i_1 \; R_k \; i_2 .$$

Define j_1 and j_2 by $i_1 \in C_{k,j_1}$ and $i_2 \in C_{k,j_2}$. Then (225) and (214) imply

(226) $$j_2 < j_1 .$$

Now, Lemma 100 implies that $i \sim_{Z_\beta} i_1$, i.e.,

(227) $$j = j_1 .$$

Also,

$$(228)\qquad \begin{cases} t' = \phi^u_\beta(x',y) \\ \quad = \Phi^u_{k,j}(x',y) \quad \text{by (222)} \\ \quad = 1-\mu^u_{k,j}(x',y)(1-\phi^u_{\beta+1}(x',y)) \quad \text{by (215)} . \end{cases}$$

Now, $t'<1$ by (224). Thus by (228)

$$(229)\qquad \mu^u_{k,j}(x',y) > 0 .$$

Therefore by (218), (226), and (227) we have

$$(230)\qquad \mu^{\hat{u}}_{k,j_2}(x,y) = 1 \quad \text{for all } \hat{u} .$$

Also, (228) implies

$$\phi^\circ_{\beta+1}(x',y) = 1 > t' \geq \phi^u_{\beta+1}(x',y),$$

so that by continuity there exists $\hat{u}$ such that

$$(231)\qquad 0 \leq \hat{u} \leq u$$

$$(232)\qquad \phi^{\hat{u}}_{\beta+1}(x',y) = t' .$$

It follows that (40) fails for $\phi^{\hat{u}}_{\beta+1}$ at (t,x,t',x',y) :

$$F(t,x,y) \text{ is below } F(t',x',y) \text{ by } (224)$$

$$t' = \phi^{\hat{u}}_{\beta+1}(x',y) \quad ((232))$$

$$0 \le t \le \phi^{u}_{\beta}(x,y) \quad \text{by } (224)$$

$$\le \phi^{\hat{u}}_{\beta}(x,y) \quad \text{by } (231)$$

$$= \phi^{\hat{u}}_{k,j_2}(x,y) \quad \text{by } (222)$$

$$= \phi^{\hat{u}}_{\beta+1}(x,y) \quad \text{by } (215) \text{ and } (230) .$$

By the choice of N, this implies $(x',y) \notin N$, so that by (217) we get a contradiction to (229).

For $(206)_\beta(\text{iv})$ we must show that ϕ^{u}_{β} satisfies (41) for $(x,y) \in p_{2,3}Y_\beta$. If $(x,y) \in p_{2,3}Y_{\beta+1}$ then this follows from $(206)_{\beta+1}(\text{iv})$ and (221). Otherwise we must have $(x,y) \in X_{k,j}$ for some k and j. So assume:

(233) $$(x,y) \in X_{k,j}$$

$$(234)\quad \begin{cases} t = \phi^u(x,y) \\ v \ \text{ is tangent to } \ I\times P\times D^s \ \text{ at } \ (t,x,y) \\ d(\phi^u\circ p_{2,3})\cdot v \geq dp_1\cdot v \\ DF\cdot v = \dfrac{\partial}{\partial t} \end{cases}$$

Statement (233) means that there exist $\tilde{x}$ in $W_\alpha^{(D_k,R_k)}(F)$ and $i \in C_{k,j}$ such that

$$(x,y) = p_{2,3}\pi_i\tilde{x} .$$

By (234) and Lemma 98 there exists $i_1 \in I(Z_\beta)$ such that

$$(t,x,y) = \pi_{i_1}\tilde{x} .$$

Lemma 100 implies that $i_1 \simeq_{z_\beta} i$, i.e. $i_1 \in C_{k,j}$. Lemma 100 also implies (by (234)) that $i_1 \in S_{Z_\beta}$, and we clearly have

$$v = \xi_{\beta,i_1}(\tilde{x}) .$$

Now (222) and (219) lead to a contradiction with (234).

It remains to prove Claim 117 (under our standing assumptions concerning F and $\phi^u_{\beta+1}$).

§III.C.5.b. PROOF of Claim 117.

Fix k, and assume that $\mu^u_{k,1},\cdots\mu^u_{k,j-1}$ have already been constructed, and that, defining $\Phi^u_{k,1},\cdots\Phi^u_{k,j-1}$ by (215), we have (216)-(220) for these smaller values of j.

Step 1. Construct a smooth function

$$P\times D^s \xrightarrow{\lambda} I$$

such that

(235) $\lambda \equiv 1$ on a neighborhood of C

(236) $\text{supp}(\lambda) \subset N$

(237) $$\text{if} \quad \tilde{x} \in W_\beta^{(D_k,R_k)}(F) \ , \ i \in C_{k,j} \ ,$$

$$i_2 \in C_{k,j_2} \ , \quad \text{and} \quad j_2<j \ , \quad \text{then}$$

$$\lambda(p_{2,3}\pi_i\tilde{x}) \neq 0 \Longrightarrow$$

$$\mu^{\hat{u}}_{k,j_2}\, p_{2,3}\pi_{i_2}\tilde{x} = 1 \quad \text{for all} \quad \hat{u} \ (0\le\hat{u}\le 1) \ .$$

This is not hard; we are merely requiring that $\lambda \equiv 0$ and $\lambda \equiv 1$ in neighborhoods of two disjoint closed sets. (For each j_2 $(1\le j_2<j)$ we have $\mu^{\hat{u}}_{k,j_2} \equiv 1$ on a neighborhood of C. It follows, since $p_{2,3}\circ\pi_{i_2}$ (for $i_2 \in C_{k,j_2}$) embeds $W_\beta^{(D_k,R_k)}$ as a closed set in $P\times D^s - C$, that $\mu^{\hat{u}}_{k,j2}\circ p_{2,3}\circ\pi_{i_2} \equiv 1$ outside of a compact set in $W_\beta^{(D_k,R_k)}$. The image of this set under $p_{2,3}\pi_i$ (for $i \in C_{k,j}$) is again compact and disjoint from C.)

Thus if we were to set $\mu^u_{k,j} = \lambda$ we would have all of (216)-(220) except for (219).

Step 2. We modify λ to obtain a $\mu^u_{k,j}$ which makes (219) true. Geometrically what we want is this: In $I\times P\times D^s$ there is a hypersurface H, the graph of $\phi^u_{k,j}$, and also a manifold $M = \pi_i W_\beta^{(D_k,R_k)}$, and there is a vector field ξ defined along M. We want that, wherever H meets M, ξ should point into the region bounded below by H. The plan for achieving this is to tilt H, without changing its intersection with M. This will require the fact that in $P\times D^s$ the vector η (image of ξ under projection) is nowhere tangent to the (embedded) image of M.

In detail: For each $i \in C_{k,j} \cap S_{Z_\beta}$, let E_i be the set where " λ fails to satisfy (219)":

$$E_i \underset{\text{def}}{=} \{(u,x,y) \in I\times P\times D^s \mid \text{ for some }$$
$$\tilde{x} \in W_\beta^{(D_k,R_k)}(F) \text{ we have } p_{2,3}\pi_i\tilde{x} = (x,y) \;,$$
$$p_1\pi_1\tilde{x} = 1 - \lambda(s,y)(1-\phi^u_{\beta+1}(x,y)) \;,$$
$$d(1-\lambda\cdot(1-\phi^u_{\beta+1}))\cdot\eta_{\beta,i} \geq (dp_1)\cdot\xi_{\beta,1} \text{ at } \tilde{x}\} \;.$$

E_i is disjoint from the closed set

$$Z \underset{\text{def}}{=} \{(u,x,y) \in I\times P\times D^s \mid 1 - \phi^u_{\beta+1}(x,y) = 0$$
$$\text{or } \lambda(x,y) = 0 \text{ or } \lambda(x,y) = 1\} \;.$$

(Let $(u,x,y) \in E_i$. If $1 - \phi^u_{\beta+1}(x,y) = 0$, whence $d(1-\phi^u_{\beta+1})(x,y) = 0$ then we have $0 \geq (dp_1)\cdot\xi_{\beta,i}$ at $\tilde{x}$ and $p_1\pi_i\tilde{x} = 1$, an impossibility by the definition of $\xi_{\beta,i}$. If $\lambda(x,y) = 0$, whence also $(d\lambda)(x,y) = 0$, then the same is again true. If $\lambda(x,y) = 1$, whence again $(d\lambda)(x,y) = 0$, then $\phi^u_{\beta+1}$ violates (41) at $\pi_i\tilde{x} = (\phi^u_{\beta+1}(x,y),x,y)$ while $(x,y) \in N$ by (236), an impossibility.) Also E_i is compact (by (235) and what was just proved), and E_i is disjoint from E_j , if $i \neq j$.

Choose open sets $\mathcal{U}_i$ in $I\times P\times D^s$ such that

(238) $$\mathcal{U}_i \supset E_i$$

(239) $$\overline{\mathcal{U}}_i \cap \overline{\mathcal{U}}_{i'} = \emptyset \quad \text{if} \quad i \neq i'$$

(240) $$\overline{\mathcal{U}}_i \cap Z = \emptyset \ .$$

Choose a smooth function

$$I\times P\times D^s \xrightarrow{g_i} I$$

such that

(241) $$g_i \equiv 1 \quad \text{in a neighborhood of} \quad E_i$$

(242) $$\operatorname{supp}(g_i) \subset \mathcal{U}_i \ .$$

Write $g_i^u(x,y) = g_i(u,x,y)$.

The function

$$\frac{d(1-\lambda(x,y)(1-\phi_{\beta+1}^u(x,y))\cdot\eta_{\beta,i} - (dp_1)\cdot\xi_{\beta,i}}{1 - \phi_{\beta+1}^u(x,y)} ,$$

defined for (u,x,y) in the compact set E_i, is bounded above by some positive number A. The image $p_{2,3}E_i \subset X_{k,j} \subset P\times D^s$ is contained in a submanifold Ω_i of $P\times D^s$, the image of an open subset of $\hat{S}^*_\beta(F)$ under $p_{2,3}\pi_i$. The vector field $\eta_{\beta,i}$ is defined along Ω_i and nowhere tangent to Ω_i. (Lemma 104). Therefore there is a smooth function

$$P\times D^s \xrightarrow{g_i'} \mathbf{R}$$

which in a neighborhood of $p_{2,3}E_i$ vanishes on Ω_i, and which satisfies

$$(dg_i')\cdot\eta_{\beta,i} > A \quad \text{on} \quad p_{2,3}E_i .$$

Choose a smooth function

$$P\times D^s \xrightarrow{g_i''} I$$

which is supported in this neighborhood and identically one on a smaller neighborhood. Choose the support of g_i'' small enough so that

$$\operatorname{supp}(g_i'') \cap X_{k,j} \subset \Omega_i$$

$$(dg_i')\cdot\eta_{\beta,i} > A \quad \text{on} \quad \mathrm{supp}(g_i'') \cap X_{k,j}$$

$$|g_i'(x,y)| \le \epsilon \quad \text{on} \quad \mathrm{supp}(g_i'') \quad ,$$

(for some $\epsilon>0$ which we have yet to choose). Setting $g_i''' = g_i' \, g_i''$, we have:

(243) $$g_i''' \equiv 0 \quad \text{on} \quad X_{k.j}$$

(244) $$(dg_i''')\cdot\eta_{\beta,i} > A \quad \text{on} \quad p_{2,3}E_i$$

(245) $$(dg_i''')\cdot\eta_{\beta,i} \ge 0 \quad \text{on} \quad X_{k,j}$$

(246) $$|g_i'''(x,y)| \le \epsilon \quad \text{for all} \quad (x,y) \in P\times D^s \quad .$$

Having done all of this for all $i \in C_{k,j} \cap I(Z_\beta)$ we set

$$\mu_{k,j}^u(x,y) = \lambda(x,y) + \sum_i g_i'''(x,y) g_i(u,x,y) \quad .$$

Thus by (240) and (242) we have

(247) $$\lambda(x,y) \in \{0,1\} \implies \mu_{k,j}^u(x,y) = \lambda(x,y) \quad .$$

We check (216)-(220).

Using (247), we obtain (216) from (235), (217) from (236), and (218) from (237).

For (220), let $(x,y) \in P\times D^s$. If $\lambda(x,y) = 0$ then we are done by (247) and (215). Likewise if $1 - \phi^u_{\beta+1}(x,y) = 0$ for all u then we are done by (215). Suppose then that $\lambda(x,y) > 0$ and (using $(206)_{\beta+1}$) that $\frac{\partial}{\partial u}(1-\phi^u_{\beta+1}(x,y)) > 0$ for all u. We will show that $\frac{\partial}{\partial u}\phi^u_{k,j}(x,y) < 0$ for all u, thus completing the proof of (220). If $(u,x,y) \notin \bigcup_i \overline{U}_i$, then near (u,x,y) we have

$$\phi^u_{k,j}(x,y) = 1 - \lambda(x,y)(1-\phi^u_{\beta+1}(x,y))$$

$$\frac{\partial}{\partial u}\phi^u_{k,j}(x,y) = - \lambda(x,y)\frac{\partial}{\partial u}(1-\phi^u_{\beta+1}(x,y)) < 0 \quad .$$

If $(u,x,y) \in \overline{U}_i$ then near (u,x,y) we have (by (239))

$$- \frac{\partial}{\partial u}\phi^u_{k,j}(x,y) = \lambda(x,y)\frac{\partial}{\partial u}(1-\phi^u_{\beta+1}(x,y))$$

$$+ g_i'''(x,y)\frac{\partial}{\partial u} \; g_i(u,x,y)(1-\phi^u_{\beta+1}(x,y)) \quad .$$

On $\overline{U}_i$ the first term is bounded below by a positive number δ, by (240) and $(206)_{\beta+1}(v)$. Therefore by (246) we are done, provided that ϵ is small enough so that

$$\epsilon \cdot \max_{(u,x,y)} \left| \frac{\partial}{\partial u} \; g_i(u,x,y)(1-\phi^u_{\beta+1}(x,y)) \right| < \delta \; .$$

Finally we check (219). Suppose that for some $\tilde{x} \in W_\beta^{(D_k,R_k)}(F)$ and $i \in C_{k,j} \cap S_{Z_\beta}$ we have

$$p_1\pi_i\tilde{x} = \Phi^u_{k,j}(p_{2,3}\pi_i\tilde{x}) \; .$$

Thus

$$p_1\pi_i\tilde{x} = |\lambda(p_{2,3}\pi_i\tilde{x})(1-\phi^u_{\beta+1}(p_{2,3}\pi_i\tilde{x}))| \quad \text{by} \quad (243) \; .$$

Let $(t,x,y) = \pi_i\tilde{x}$.

<u>Case 1</u>. $(u,x,y) \notin E_i$. Then by definition of E_i

$$d(1-\lambda\cdot(1-\phi^u_{\beta+1}))\cdot\eta_{\beta,i}(\tilde{x}) < (dp_1)\cdot\xi_{\beta,i}(\tilde{x}) \quad .$$

On the other hand,

$$d(g_i'''\cdot g_i^u\cdot(1-\phi^u_{\beta+1}))\cdot\eta_{\beta,i}(\tilde{x}) =$$

$$(g_i(u,x,y)\cdot(1-\phi^u_{\beta+1}(x,y))) \;\; (dg_i''')\cdot\eta_{\beta,i}(\tilde{x}) \; , \quad \text{by} \quad (243)$$

$$\geq 0 \quad \text{by } (245)$$

Since by (242) we have in a neighborhood of $p_{2,3}\pi_i\tilde{x}$

$$\text{(249)} \qquad \mu^u_{k,j}(x,y) = \lambda(x,y) + g_i'''(x,y)g_i(u,x,y),$$

the desired conclusion follows:

$$(d\Phi^u_{k,j})\cdot\eta_{\beta,i}(\tilde{x}) < (dp_1)\cdot\xi_{\beta,i}(\tilde{x}) .$$

<u>Case 2</u>. $(u,x,y) \in E_i$. Since (249) is again valid in a neighborhood of $p_{2,3}\pi_i\tilde{x}$, and (248) is also true, the result this time follows from (241), (244), and the choice of A.

There is one other small matter to check: $\mu^u_{k,j}$ must take values in I. By (242), (246), and the fact that g_i takes values in I, it will suffice to choose ϵ such that in $\overline{\mathcal{U}}_i$ λ takes values between ϵ and $1-\epsilon$. This is possible by (240).

This completes the proof of Claim 117, and thus the proof that $(206)_{\beta+1} \Longrightarrow (206)_\beta$, and thus the proof of $(206)_0$, and thus the proof of Claim 116, and thus the proof of Claim 113, and thus the proof that $(189)_{\alpha+1} \Longrightarrow (189)_\alpha$, and thus the proof of $(189)_0$.

§III.D. <u>One Last Sunny Collapse</u>.

Having proved $(189)_0$ we can easily complete the proof of Theorem D. Let $F = (h,f,p_3)$ and $\{\mathcal{O}_j\}_{j=1}^{a}$ satisfy the conditions of $(204)_0$. Consider the closed set

$$X \underset{\text{def}}{=} p_{2,3}Y_0 \cup (\partial P\times D^s) \subset P\times D^s$$

By assumption X satisfies

(250) $$f(I\times X \cap I\times P\times\overline{\mathcal{O}}_j) \cap Q_j = \emptyset \quad \text{for} \quad 1\le j\le a .$$

Set

$$Y \underset{\text{def}}{=} \bigcup_{j=1}^{a} \left(f^{-1}(Q_j) \cap I\times P\times\overline{\mathcal{O}}_j \right) ,$$

a closed subset of $I\times P\times D^s$. Statement (250) implies

(251) $$(t,x,y) \in Y \implies (x,y) \notin X .$$

Also, by the definition of fibered concordance

(252) $$(t,x,y) \in Y \implies t>0 .$$

It follows easily from (251) and (252) and the compactness of X and Y that there is a smooth function

$$P \times D^s \xrightarrow{\phi} (0,1]$$

such that

(253) $\phi(x,y) < t \quad \text{for} \quad (t,x,y) \in Y$

(254) $\phi(x,y) = 1 \quad \text{for} \quad (x,y) \in X \ .$

Because of (254), ϕ satisfies (49), (50), and (51):

(49) is immediate.

For (50), suppose

$$F(t,x,y) \text{ is below } F(t',x',y) \ .$$

Then either $x \in \partial P$ or $(t,x,t',x',y) \in S_0(F)$. In either case $(x,y) \in X$, so that $\phi(x,y) = 1$. But $t<1$ because

$$h(t,x,y) < h(t',x',y) \leq 1 \ .$$

Thus $\phi(x,y) > t$.

For (51), suppose (t_0,x_0,y_0) and v satisfy

$$\phi(x_0,y_0) \le t_0 < 1$$

$$(DF)\cdot v = \frac{\partial}{\partial t} .$$

Either $x_0 \in \partial P$ or $(t_0,x_0,y_0) \in S_1(F)$. In either case $(x_0,y_0) \in X$, so that $\phi(x_0,y_0) = 1$, a contradiction.

Thus by Lemma 30

$$\phi^u \underset{\text{def}}{=} 1 - u(1-\phi)$$

is a sunny collapse. Let

$$F^u = (h^u,f^u,p_3)$$

be the associated isotopy. F^u preserves (187) by (49) (exactly as in the proof of Claim 115). Therefore F^1 is a new representative for the isotopy class. By (48), for any $y \in \overline{\mathcal{O}}_j$ we have

$$\begin{aligned} f^1(t,x,y) \in Q_j \quad &\Longrightarrow \quad f^0(t\phi(x,y),x,y) \in Q_j \\ &\Longrightarrow \quad (t\phi(x,y),x,y) \in Y \\ &\Longrightarrow \quad \phi(x,y) < t\phi(x,y) \le \phi(x,y) \end{aligned}$$

by (253), a contradiction. Thus F^1 satisfies (188), and Theorem D is proved.

BIBLIOGRAPHY

[A] J. F. Adams, Algebraic Topology - A Student's Guide, Cambridge University Press, 1972.

[BCCGHM] M. Bökstedt, G. Carlsson, R. Cohen, T. Goodwillie, W.-c Hsiang, I. Madsen, "The Algebraic K-Theory of Simply-Connected Spaces", in preparation.

[BL] D. Burghelea and R. Lashof, "Stability of Concordances and the Suspension Homomorphism", Annals of Math. 105 (1977), 449-472.

[BLR] D. Burghelea, R. Lashof, M. Rothenberg, Groups of Automorphisms of Manifolds, (Lect. Notes in Math. No. 473) Springer-Verlag, 1975.

[BW] M. G. Barratt and J. H. C. Whitehead, "The First Non-vanishing Group of an (n+1)-Ad ", Proc. London Math. Soc. (3) 6 (1956), 417-439.

[CCGH] G. Carlsson, R. Cohen, T. Goodwillie, W.-c. Hsiang, "The Free Loop Space and the Algebraic K-Theory of Spaces", K-Theory 1 (1987), 53-82.

[DHS1 W. Dwyer, W.-C. Hsiang, R. Staffeldt, "Pseudo-isotopy and Invariant Theory - I", Topology, 19 (1980), 367-385.

[DHS2] W. Dwyer, W.-C. Hsiang, R. Staffeldt, "Pseudo-isotopy and Invariant Theory - II: Rational Algebraic K-Theory of a Space with Finite Fundamental Group", Lect. Notes in Math. 778 (1980), 418-441.

[Gl] G. Glaeser, "L'Interpolation des Fonctions Différentiables de Plusieurs Variables", Proceedings of Liverpool Singularities Symposium II, Lecture Notes in Math. 209, Springer-Verlag, 1971, 1-33.

[Go1] T. Goodwillie, "Calculus I: The First Derivative of a Homotopy Functor", preprint.

[Go2] T. Goodwillie, "Calculus II: The Taylor Series of a Homotopy Functor", in preparation.

[Go3] T. Goodwillie, "Relative Algebraic K-Theory and Cyclic Homology", Annals of Math 124 (1986), 347-402.

[Har] R. Hartshorne, Algebraic Geometry, (Graduate Texts in Mathematics 52), Springer-Verlag, 1977.

[Hat] A. Hatcher, "Concordance Spaces, Higher Simple Homotopy Theory, and Applications", Proc. Symp. Pure Math. 32, Part I (1978), 3-22.

[Hi] M. Hirsch, Differential Topology, Springer-Verlag, 1976.

[HS1] W.-C. Hsiang and R. E. Staffeldt, "A Model for Computing Rational Algebraic K-theory of Simply Connected Spaces", Invent. Math. 68 (1982), 227-239.

[HS2] W.-C. Hsiang and R. E. Staffeldt, "Rational Algebraic K-Theory of a Product of Eilenberg-Maclane Spaces", Contemp. Math. (1983), 95-114

[Hu] J. F. P. Hudson, "Embeddings of Bounded Manifolds", Proc. Camb. Phil. Soc. 72 (1972), 11-25.

[HW] A. Hatcher and J. Wagoner, "Pseudoisotopies of Compact Manifolds", Asterisque 6 (1973), Soc. Math. de France, Paris.

[I_1] K. Igusa, "What Happens to Hatcher and Wagoner's Formula for $\pi_0 C(M)$ When the First Postnikov Invariant of M is Nontrivial?", Lect. Notes in Math. 1046 (1984), 104-172.

[I_2] K. Igusa, "The $Wh_3(\pi)$ Obstruction for Pseudoisotopy". Thesis, Princeton Univ., 1979.

[Ma1] J. N. Mather, "Stability of C^∞ Mappings: II. Infinitesimal Stability Implies Stability", Annals of Math. 89 (1969), 254-291.

[Ma2] J. N. Mather, "Stability of C^∞ Mappings: V, Transversality", Advances in Math. 4 (1970), 301-336.

[Mi1] K. C. Millett, "Piecewise Linear Concordances and Isotopies", Memoirs of the American Math. Soc. 153, A.M.S. 1974.

[Mi2] K. C. Millett, "Piecewise Linear Embeddings of Manifolds", Illinois J. Math. 19 (1975), 354-369.

[O] C. Ogle, "A Generalized Trace Map for Waldhausen's Algebraic K-theory of Spaces, and Applications, preprint (Ohio State).

[S] I. R. Shafarevich, Basic Algebraic Geometry, Springer-Verlag, 1974.

[W1] F. Waldhausen, "Algebraic K-Theory of Spaces", Lect. Notes in Math. 1126 (1985), 318-419.

[W2] F. Waldhausen, "Algebraic K-Theory of Spaces, Concordance, and Stable Homotopy Theory" in W. Browder (ed.) Annals of Math Studies 113 (1987), 392-417.

General instructions to authors for

PREPARING REPRODUCTION COPY FOR MEMOIRS

For more detailed instructions send for AMS booklet, "A Guide for Authors of Memoirs." Write to Editorial Offices, American Mathematical Society, P.O. Box 6248, Providence, R.I. 02940.

MEMOIRS are printed by photo-offset from camera copy fully prepared by the author. This means that, except for a reduction in size of 20 to 30%, the finished book will look exactly like the copy submitted. Thus the author will want to use a good quality typewriter with a new, medium-inked black ribbon, and submit clean copy on the appropriate model paper.

Model Paper, provided at no cost by the AMS, is paper marked with blue lines that confine the copy to the appropriate size. Author should specify, when ordering, whether typewriter to be used has **PICA**-size (10 characters to the inch) or **ELITE**-size type (12 characters to the inch).

Line Spacing — For best appearance, and economy, a typewriter equipped with a half-space ratchet — 12 notches to the inch — should be used. (This may be purchased and attached at small cost.) Three notches make the desired spacing, which is equivalent to 1-1/2 ordinary single spaces. Where copy has a great many subscripts and superscripts, however, double spacing should be used.

Special Characters may be filled in carefully freehand, using dense black ink, or **INSTANT** ("rub-on") **LETTERING** may be used. AMS has a sheet of several hundred most-used symbols and letters which may be purchased for $5.

Diagrams may be drawn in black ink either directly on the model sheet, or on a separate sheet and pasted with rubber cement into spaces left for them in the text. Ballpoint pen is not acceptable.

Page Headings (Running Heads) should be centered, in CAPITAL LETTERS (preferably), at the top of the page — just above the blue line and touching it.

LEFT-hand, EVEN-numbered pages should be headed with the AUTHOR'S NAME;

RIGHT-hand, ODD-numbered pages should be headed with the TITLE of the paper (in shortened form if necessary).

Exceptions: PAGE 1 and any other page that carries a display title require NO RUNNING HEADS.

Page Numbers should be at the top of the page, on the same line with the running heads.

LEFT-hand, EVEN numbers — flush with left margin;

RIGHT-hand, ODD numbers — flush with right margin.

Exceptions: PAGE 1 and any other page that carries a display title should have page number, centered below the text, on blue line provided.

FRONT MATTER PAGES should be numbered with Roman numerals (lower case), positioned below text in same manner as described above.

MEMOIRS FORMAT

It is suggested that the material be arranged in pages as indicated below.
Note: Starred items (*) are requirements of publication.

Front Matter (first pages in book, preceding main body of text).

Page i — *Title, *Author's name.

Page iii — Table of contents.

Page iv — *Abstract (at least 1 sentence and at most 300 words).

Key words and phrases, if desired. (A list which covers the content of the paper adequately enough to be useful for an information retrieval system.)

**1980 Mathematics Subject Classification (1985 Revision)*. This classification represents the primary and secondary subjects of the paper, and the scheme can be found in Annual Subject Indexes of MATHEMATICAL REVIEWS beginnning in 1984.

Page 1 — Preface, introduction, or any other matter not belonging in body of text.

Footnotes: *Received by the editor date.
Support information — grants, credits, etc.

First Page Following Introduction - Chapter Title (dropped 1 inch from top line, and centered). Beginning of Text.

Last Page (at bottom) - Author's affiliation.